MUNICIPAL WASTE TREATMENT TECHNOLOGY

城市垃圾处理技术

杜海霞（Haixia Du） 吴慧芳（Huifang Wu） 编

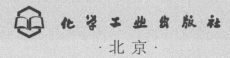

化学工业出版社

·北京·

Synopsis

The textbook is designed for the management, treatment and disposal of municipal waste. The content contains types and characteristics of waste; waste storage, segregation and collection; status of municipal waste treatment and resource recovery; analysis and determination of municipal solid waste; landfill treatment for municipal solid waste; thermal treatment for municipal solid waste; anaerobic digestion for municipal solid waste; composting for municipal solid waste; and life cycle assessment (LCA) for municipal solid waste management.

This textbook focuses on the introduction of classification, management and main treatment technologies of municipal waste. It involves the advanced concept of classification and management of municipal waste and expounds the principle and application of major treatment and resource-recycling technologies such as incineration, anaerobic digestion and composting. The textbook provides professional knowledge of municipal waste treatment and resource recycling for the design and technology application of urban environmental management and comprehensive environmental treatment project.

图书在版编目（CIP）数据

城市垃圾处理技术＝MUNICIPAL WASTE TREATMENT TECHNOLOGY：英文/杜海霞，吴慧芳编．—北京：化学工业出版社，2020.9（2023.7重印）
　ISBN 978-7-122-37006-8

Ⅰ.①城… Ⅱ.①杜… ②吴… Ⅲ.①城市-垃圾处理-英文 Ⅳ.①X799.305

中国版本图书馆 CIP 数据核字（2020）第 085141 号

责任编辑：满悦芝　　　　　　　　　　　　文字编辑：杨振美　陈小滔
责任校对：王素芹　　　　　　　　　　　　装帧设计：张　辉

出版发行：化学工业出版社（北京市东城区青年湖南街13号　邮政编码100011）
印　　装：涿州市般润文化传播有限公司
710mm×1000mm　1/16　印张10　字数166千字　2023年7月北京第1版第4次印刷

购书咨询：010-64518888　　　　　　　　　售后服务：010-64518899
网　　址：http://www.cip.com.cn
凡购买本书，如有缺损质量问题，本社销售中心负责调换。

定　价：49.80元　　　　　　　　　　　　　　　　　　　版权所有　违者必究

Preface

Urbanization and rapid economic development lead to a sharp increase on the generation of municipal waste. The effective treatment of municipal waste and its resource recovery are of great significance to the realization of a healthy, fast and suitable urban living environment, the reduction of environmental pollution and the construction of a resource recyclable society.

Based on the current scarcity of English teaching materials for municipal waste treatment and resource recovery, this textbook is designed for setting up English teaching course of municipal waste treatment, in order to increase students' English learning atmosphere. It emphasizes that students fully understand the English teaching content of municipal waste treatment technology, and meanwhile promotes students' strong interest in English teaching course.

With a gradual structure, the textbook follows the cognitive rules of students. It focuses on the classification, management, resource-based treatment technology of municipal waste and design of a life cycle assessment for estimating effect of municipal waste on environment and human health. The textbook introduces several advanced waste management methods and treatment technologies, which provide reference and guidance for the understanding and mastering of comprehensive treatment and disposal of municipal waste and its efficient management.

At the same time, the textbook is written in English to provide reference and guidance for English teaching, in order to improve the international competitiveness of students in cultural literacy, professional skills and other aspects.

Haixia Du
Huifang Wu
2020. 5

Contents

Chapter 1　Introduction   **1**

    1.1　Types and characteristics of wastes　3

    1.2　Waste prevention　9

Chapter 2　Waste storage, segregation and collection   **16**

    2.1　Source segregation　17

    2.2　Waste storage　18

    2.3　Waste collection　19

    2.4　Waste separation　22

Chapter 3　Status of municipal waste treatment and resource recovery   **24**

    3.1　Status of municipal waste treatment　24

    3.2　Resource recovery　26

Chapter 4　Analysis and determination of municipal solid waste   **29**

    4.1　Physical properties of municipal solid waste　30

    4.2　Chemical properties of municipal solid waste　33

    4.3　Biological properties of municipal solid waste　40

Chapter 5　Landfill treatment for municipal solid waste   **45**

    5.1　What is landfill?　45

    5.2　Types of landfills　46

    5.3　Operations of landfills　47

5.4　Advantages of landfills　48

5.5　Social and environmental impact of landfills　48

5.6　Landfill gas　49

5.7　Landfill leachate production, collection and management　53

Chapter 6　Thermal treatment for municipal solid waste　58

6.1　Incineration　61

6.2　Pyrolysis　76

6.3　Gasification　83

Chapter 7　Anaerobic digestion for municipal solid waste　95

7.1　Description　96

7.2　Process　98

Chapter 8　Composting for municipal solid waste　118

8.1　Terminology　119

8.2　Fundamentals　120

8.3　Materials that can be composted　123

8.4　Uses　125

8.5　Composting technologies　126

8.6　Regulations　129

8.7　Examples　130

8.8　History　131

Chapter 9　Life cycle assessment (LCA) for municipal solid waste management　133

9.1　Introduction—What is LCA and how is it useful?　134

9.2　Goals and purpose of LCA　135

9.3　Four main phases　136

9.4　LCA uses　140

9.5   Data analysis   141

9.6   Variants   143

9.7   Exergy based LCA   146

9.8   Life cycle energy analysis   146

9.9   Critiques   149

9.10 Supporting waste management decisions-Examples   151

References 152

Chapter 1 Introduction

With the increase in population and increasing resource utilization patterns, in fact we have surpassed the carrying capacity of the planet. The earth's environmental assets are now insufficient to sustain our demands and economic activities. The global warming has underscored the danger of overstepping the earth's ability to absorb our waste products. The earth's ability to absorb our waste is a major factor influencing the adaptation of the waste treatment technology. Landfilling is probably the oldest organized waste management technology. Until 1970s, landfilling has been carried out as an unceremonious dumping of waste at any convenient location without considering health, safety, environmental protection, or cost efficiency. Now, in the urban area, availability of space for landfilling is becoming scarce and a very serious issue. Currently, the trend in the countries having advanced waste management system is to reduce the wastes ending up in landfills. Two decades ago, only about 30% of the municipal solid wastes produced were recycled. In the last decade, the importance of waste reduction and recycling was realized and through extensive education program on the Source Separation of Domestic Waste, a recovery rate of 48% was achieved in 2012.

In addition to recycling, the increasing waste-to-energy programs and the developments in the technology and pollution control devices further reduced the quantities reaching landfills mainly in Europe while it could be a blueprint for other countries in the future. This is especially applicable to the regions where finding new landfill space is a challenge and those regions where these technologies are not yet fully implemented. It may also be expected that improved and more sustainable product design would be available in the near future that would change the face of the energy recovery systems. Despite the oddities, landfilling is unavoidable and often the final inert fraction needs to be buried. The landfill design, operation and management is continuously being researched and new technologies are inducted to avoid the air and water pollution. For example, the landfill gas collection provides scope for reducing the green-house gas (GHG). But the economics of the landfill collection and energy recovery need to be demonstrated convincingly. Because the average methane content of landfill gas is around 50% due to partial oxidation in the landfill and part of the gas produced in landfills is lost to the atmosphere even with an effective gas collection system, this low level of methane in the landfill gas necessitates critical upgrading works jeopardizing the advantages of the landfill gas extraction.

Integrated waste management is a concept for designing and implementing new waste management systems and for analyzing and optimizing existing systems. In this concept, both technical and non-technical components of the management system should be analyzed together. Currently, with the development of new policies, regulations and waste management business, non-technical components including public engagement and education are unavoidable and key to the successful implementation of many recycling and recovery programs. As the modern waste management hierarchy insists, public participation and change of public's perception are the first step to achieve the waste reduction while recycling and reuse also require the support of technology. Therefore, a modern integrated waste management is the need of the time while the sustainability must be infused in all components considering the supply and demand of the resources.

The municipal solid waste industry has four components: recycling, composting, disposal, and waste-to-energy via incineration. There is no single approach that can be applied to the management of all waste streams, therefore the Environmental Protection Agency, a

U. S. federal government agency, developed a hierarchy ranking strategy for municipal solid waste. The Waste Management Hierarchy is made up of four levels ordered from most preferred to least preferred methods based on their environmental soundness: Source reduction and reuse; recycling or composting; energy recovery; treatment and disposal, as shown in Fig. 1.

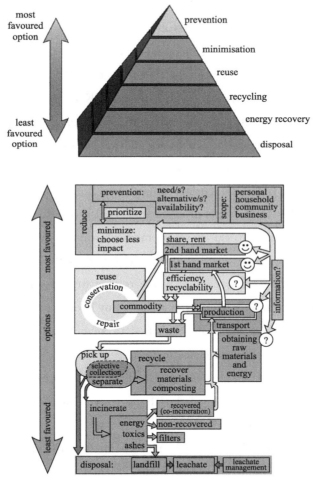

Fig. 1 Waste management hierarchy for sustainable use of resources
Source: Environment Bureau of Hong Kong

1.1 Types and characteristics of wastes

Waste can be classified depending on its source. Classification of waste is also determined by the socio-economic sector generating the waste. The types of wastes include: (1) municipal solid waste; (2) industrial waste; (3) hazardous waste; (4) agricultural waste; (5) medical waste; (6) sludge; (7) chemical waste; (8) biological waste; (9) E-waste.

The composition of municipal solid waste varies greatly from municipality to municipality, and it changes significantly with time. In municipalities which have a well developed waste recycling system, the waste stream mainly consists of intractable wastes such as plastic film and non-recyclable packaging materials. At the start of the 20th century, the majority of domestic waste (53%) in the UK consisted of coal ash from open fires. In developed areas without significant recycling activity it predominantly includes food wastes, market wastes, yard wastes, plastic containers and product packaging materials, and other miscellaneous solid wastes from residential, commercial, institutional, and industrial sources. Most definitions of municipal solid waste do not include industrial wastes, agricultural wastes, medical wastes, radioactive wastes or sewage sludge. Waste collection is performed by the municipality within a given area. The term residual waste relates to waste left from household sources containing materials that have not been separated out or sent for reprocessing. Waste can be classified in several ways but the following list represents a typical classification:

(1) Biodegradable wastes: food and kitchen waste, green waste, paper (most can be recycled although some plant material difficult to compost may be excluded);

(2) Recyclable materials: paper, cardboard, glass, bottles, jars, tin cans, aluminum cans, aluminum foil, metals, certain plastics, fabrics, clothes, tires, batteries, etc. ;

(3) Inert wastes: construction and demolition waste, dirt, rocks, debris;

(4) Waste electrical and electronic equipment (WEEE): electrical appliances, light bulbs, washing machines, TVs, computers, screens, mo-

bile phones, alarm clocks, watches, etc. ;

(5) Composite wastes: waste clothing, Tetra Packs, waste plastics such as toys;

(6) Hazardous wastes: including most paints, chemicals, tires, batteries, light bulbs, electrical appliances, fluorescent lamps, aerosol spray cans, and fertilizers;

(7) Toxic wastes: including pesticides, herbicides, and fungicides;

(8) Biomedical wastes, expired pharmaceutical drugs, etc.

1.1.1 Municipal solid waste

Municipal solid waste (MSW) includes durable goods, non-durable goods, containers and packaging materials, food wastes and yard trimmings, and miscellaneous inorganic wastes. Sometime it is also referred as garbage (e. g., milk cartons and coffee grounds), refuse (e. g., metal scrap, wallboard, and empty containers), sludge from waste treatment plants, water supply treatment plants, or air pollution control facilities (e. g., scrubber sludge), other discarded materials, including solid, semi-solid, liquid, or contained gaseous material resulting from industrial, commercial, mining, agricultural, and community activities (e. g., boiler slag or fly ash). MSW is mainly generated from households and wastes of similar character from shops, markets and offices, open areas, and treatment plant sites. MSW mainly consists of biodegradable, non-biodegradable and inert components. The physiological composition of MSW from different continents is presented in Table 1. As observed from Table 1, it can be concluded that organic waste component is one of the challenging wastes that needs proper treatment and disposal. MSW also comprises of non-biodegradable and inert components.

Table 1 Composition of MSW from different continents

Continent	Organic/%	Paper/%	Plastic/%	Glass/%	Metal/%	Others/%
Europe	35	21	10	6	4	23
Asia	51	14	11	4	4	16

Continued

Continent	Organic/%	Paper/%	Plastic/%	Glass/%	Metal/%	Others/%
South America	51	15	11	3	3	18
North America	42	22	13	3	6	14
Australia	52	18	8	5	4	13
Africa	56	9	10	3	3	20

Source: *World Bank* (2010).

1.1.2 Industrial waste

Industrial waste is defined as waste generated by manufacturing or industrial processes. It includes biodegradable waste, dirt and gravel, masonry and concrete, scrap metals, trash, oil, solvents, chemicals, weed grass and trees, wood and scrap lumber, and similar wastes. Industrial waste may be solid, liquid or gases. It could be further classified into hazardous and non-hazardous waste. Non-hazardous industrial wastes are those that do not meet the EPA's definition of hazardous waste and are not municipal waste. Thus, it could be stated that non-hazardous industrial wastes are in between the characteristics of MSW and hazardous waste. These wastes may be toxic, ignitable, corrosive or reactive. If improperly managed, this waste can pose dangerous health and environmental consequences.

1.1.3 Hazardous waste

U.S. EPA (2005) defined hazardous waste as a waste, or combination of wastes, which because of its quantity, concentration, or physical, chemical, or infectious characteristics may: (1) cause, or significantly contribute to, an increase in mortality or an increase in serious irreversible, or incapacitating reversible, illness; or (2) pose a substantial threat or potential hazard to human health or the environment when improperly treated, stored, transported, or disposed of, or otherwise managed. U.S. EPA has recommended four characteristics as the means for identifying hazardous waste: ignitability, corrosivity, reactivity, and toxicity.

Ignitable wastes can readily catch fire and lead to combustion. Paints, cleaners, and other industrial wastes pose such fire hazard. Most ignitable wastes are liquid in physical form.

Corrosive wastes are acidic or alkaline (basic) wastes, which can readily corrode or dissolve flesh, metal, or other materials. They are also among the most common hazardous wastes. Waste sulphuric acid from automotive batteries is an example of a corrosive waste.

A reactive waste is one that readily explodes or undergoes violent reactions. Waste is considered reactive if it meets any of the following criteria:

(1) it can explode or violently react when exposed to water, when heated, or under normal handling conditions;

(2) it can create toxic fumes or gases when exposed to water or under normal handling conditions;

(3) it generates toxic levels of sulfide or cyanide gas when exposed to a pH range of $2 \sim 12.5$.

The leaching of toxic compounds or elements into groundwater or drinking water supplies from wastes disposed off in landfills is one of the most common ways to identify toxicity of the chemical found in industrial wastes.

1.1.4 Agricultural waste

Agricultural waste is any substance or object from premises used for agriculture or horticulture, which the holder discards, intends to discard or is required to discard. It is waste specifically generated by agricultural activities. Agricultural waste consist of almond, baggage, straw, corn, cotton, peanut, rice hulls, straw and wheat straw. They have very high heating or calorific value. Other chemical component includes ash content, carbon, oxygen and nitrogen.

1.1.5 Medical waste

Medical waste is also known as infectious waste or biomedical waste. It is defined as waste generated during diagnosis, testing, treatment, research or production of biological products for humans or animals. It includes syringes, live vaccines, laboratory samples, body parts, bodily fluids and waste, sharp needles, cultures and lancets. The main sources are hospitals, medical clinics and laboratories. This waste can be detrimental to human health hence public and individuals must follow procedures not to come in contact with this kind of waste. The categories and types of medical waste includes human anatomical waste, animal waste, microbiological and biotechnology waste, waste sharps, discarded medicines, liquid wastes, etc.

1.1.6 E-Waste

E-waste is the term, which is commonly used to describe "Waste Electrical and Electronic Equipment (WEEE)". Informal processing of E-waste in developing countries may cause serious pollution and health problems, though these countries are also most likely to reuse and repair electronics. All electronic scrap components, such as CRTs, may contain contaminants such as lead, cadmium, beryllium, or brominated flame-retardants. Even in developed countries recycling and disposal of E-waste may involve significant risk to workers and communities, and great care must be taken to avoid unsafe exposure in recycling operation and leaching of material such as heavy metals from landfills and incinerator ashes. The sources and types of E-waste are given in Table 2.

Table 2　Sources and types of E-waste

Sources	Types
Household appliances	Washing machines, dryers, refrigerators, irons, TVs, monitors, toasters, coffee machines, air conditioners
Offices, information and communication centers	PCs, laptops, mobiles, telephones, fax machines, copiers, printers, Xerox machines
Entertainment and consumer electronics	Televisions, VCRs, DVD players, CD players, DVDs, CDs, radios, cassettes, Hi-Fi sets
Lighting equipments	Fluorescent tubes, sodium lamps, used bulbs, led lights, CFLs, tube lights

Continued

Sources	Types
Electric and electronic tools	Drills, electric saws, sewing machines, lawn movers, electric motors, fans
Toys and sports materials	Electric toy trains, cars, bikes, remote controls, coin slot machines, treadmills, etc.
Medical instruments and equipments	
Surveillance and control equipments	CCTV cameras, monitors
Automatic issuing machines	

Source: Lim and Schoenung (2010); Bouvier and Wagner (2011).

1.1.7 Generation of waste

Generation of waste is directly proportional to population growth. Generation of MSW is increasing with leaps and bounds. Generation of industrial waste and hazardous waste depends on production rate and demand in the market. As described by the World Bank (2010), the developed countries are the main producers of waste. They generate about 46% of total global waste. This is mainly due to the behavioral and cultural aspect of the people residing in developed countries.

It could also be estimated that most of the waste generated in high income countries are treated properly. This does not hamper the progress of the nation as compared with other low, middle and upper middle income countries. MSW generation also depends on cultural and behavioral aspect of individuals as well.

1.2 Waste prevention

1.2.1 What is waste prevention?

According to the revised European Union (EU) Directive on Waste (2008-98-EC), widely known as Waste Framework Directive (WFD-Article 3, clause 12 & 13), waste prevention encompasses complex actions and measures, and a range of policy options in a broader sense, taken before a material or a product is characterized as waste. These actions and measures ought to reduce: (1) the amount of waste, through a Paradigm shift in the production and consumption patterns, including the re-design of products, their reuse or the extension of their life span; (2) the adverse impact of the generated waste on both the environment and human health; and (3) the content of harmful substances (i.e. the toxicity) in materials and products.

Based on the above definition, waste prevention can be achieved through three different but often intersecting pathways (Fig. 2): (1) the strict avoidance of waste; (2) the diversion of material and product flows from disposal; (3) and the elimination of waste toxicity. The first two paths concern the quantitative side of waste prevention, while the third concerns the qualitative. In the first pathway, waste prevention targets waste generation at its source, exploiting policies that limit unnecessary consumption and leading to the design and/or the consumption of products that generate less waste. The second approach, refers to materials and products that approach or have already reached their end-of-life, but

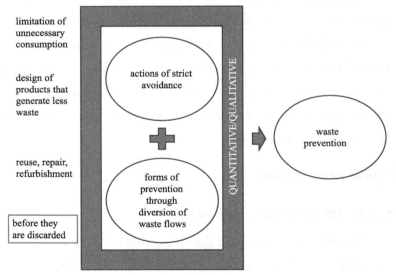

Fig. 2 The three pathways of waste prevention: strict avoidance of waste, diversion of material and product flows from disposal, and elimination of waste toxicity

are not yet discarded. It includes actions that extend the useful life or encourage the reuse of products. The third path, the elimination of waste toxicity, touches on the qualitative side of waste prevention. It is widely known that the contemporary production chain uses a large variety of synthetic materials that are not found in natural ecosystems and poses enormous negative impact on the environment. The reduction of the hazardous content of waste will limit the human and environmental exposure to hazardous substances.

A relatively different approach to the concept of waste prevention has been introduced by the Organization for Economic Co-operation and Development (OECD), seeing it as part of waste minimization (Fig. 3). The latter embraces all the material saving and recovery options, which can be applied before and after a material/product has been labeled as waste. Referring to a report of EEA (2002), *Salhofer et al.* (2008) indicated that waste minimization includes preventive measures (such as prevention, reduction at source and reuse of products) as well as some other waste management procedures aiming at quality improvements and recycling.

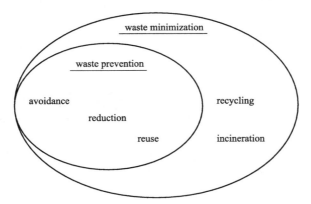

Fig. 3　An illustration of waste prevention in the context of integrated waste management, based on the definition of the OECD (2002)

Yet, approaches from both the OECD and the WFD have two important characteristics in common: (1) the use of a "life-cycle thinking" on waste management policies, and (2) the placement of waste prevention within the framework of waste management.

A basic understanding of two concepts is essential for approaching waste prevention.

The first is to realize that actions for waste prevention can be taken at all the steps in a product's life cycle, from production and supply to consumption and discard (Fig. 4). Various factors play a key role in each step and different policy tools are best suited to promote prevention. In all the steps, policies and actions for waste prevention typically focus on priority waste streams, i. e. those waste streams for which, due to their quantity and environmental footprint, prevention can most effectively reduce environmental impacts and achieve quantifiable environmental benefits. For some waste streams large differences can be achieved at the production and supply stage, while others are better suited for actions at the consumption and discard level. For most of waste streams, however, significant prevention potential may be identified at all stages of the product's life cycle.

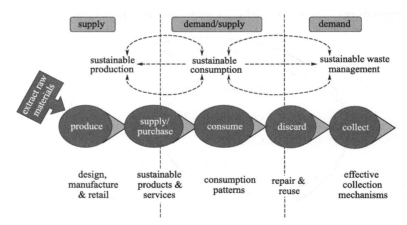

Fig. 4　Waste prevention refers to all the steps in a product's life cycle

Source: DEFRA (2009).

The second concept concerns the barriers of waste prevention implementation. Although waste prevention has been placed at the top of the waste management hierarchy in many national and international policy frameworks, waste prevention policies have not brought the expected significant results so far. Mainly, the cause rests on measurement limitations and therefore the lack of specific quantitative targets, conflict of interests, inconsistent definitions, the lack of comprehensive strategies and the human attitudes and environmental awareness.

1.2.2 Waste prevention in the context of sustainability

Waste management influences a range of broader concerns such as energy and material security, climate change, environmental protection, economy, justice and equity, which are all directly related to the three pillars of sustainability, i.e. environmental protection, economic development and social justice. Waste prevention is a useful multipurpose tool for the consolidation of sustainability, as it brings about direct or indirect ways of tackling: (1) the growing-quantitative and qualitative-waste burden; (2) the resource-intensive production and consumption patterns; and (3) the waste management costs.

In the short and long term, waste prevention furthers the efficient use of resources and can be perceived as a substantial element for fostering sustainable environmental management. Therefore, it is not surprising that quite often waste prevention is integrated in national sustainable development programs or policies. Waste prevention opportunities occur at every stage of the life cycle of a material or product; therefore, it may play a significant role in relation to the reduction of waste generation and the overall waste management planning, as well as to other upstream steps of the production line. It may also contribute to the restriction of the (over) exploitation of resources by leveraging their use.

The contribution of waste prevention actions to the abatement of global climate change could be valuable, since the GHG emission of the waste sector accounts for approximately 5% of the global greenhouse budget. According to *Yusuf et al.* (2012), the main sources of anthropogenic methane emissions belong to three sectors, namely agriculture (53%), energy (28%) and waste (19%). However, if the overall impact of resources extraction and use is considered, the impact of inefficient use of resources/waste production on climate change is considerably larger, albeit difficult to estimate at an acceptable level of accuracy.

An important aspect of waste prevention refers to food waste, especially their avoidable fraction. Hunger still remains a serious problem in the 21st century, and food waste disposal is a significant ethical and environmental issue. In EU-MS alone, 79 million citizens live below the poverty line, while approximately 89 Mt (amounting to 179 kg per capita) of food waste is disposed annually (EC 2011). Ethical issue, more or less directly related to waste prevention, is the concern about environmental equity when addressing legal and illegal shipments of waste in Less Developed Countries (LDC). Even in the legal shipment, additional measures must be taken to ensure that waste is treated in a safe way for the well being of workers and the inhabitants.

1.2.3 Summary

Since waste prevention was placed at the top of the waste management hierarchy and, therefore, brought to the forefront of discussions on environmental security and sustainable development, its relation to the whole resources management chain has been intensively investigated.

The relevant discourse and a multitude of studies indicate that waste prevention activities can-and should-be taken at all steps of the product's life chain. Waste prevention should be considered a shared priority of central and local authorities, the private sector and individuals/consumers. At each step of a product's life cycle, attention is paid on certain priority waste streams, in order to minimize the negative environmental impacts while maximizing the possible environmental, economic and social benefits.

Effective actions are underway by international organizations, central and local authorities in various countries, as well as the private sector, setting precedents for worldwide adaptation and replication. Among them, some initiatives focus on restraining unsustainable consumption patterns. Yet, much more work needs to be done, engaging the main principles of sustainable production, supply, consumption and discard patterns (e. g. resource efficiency, dematerialization, eco-design, lifestyle) and waste management. Moreover, there are still several barriers and challenges, such as methodological limitations, conflicts of interest and established human attitudes and behaviors that need to be resolved. An effort to approach waste prevention targets, considering social and market patterns, might hold a key to the design of effective waste prevention measures and the move to the-so needed-resource efficient societies.

Chapter 2 Waste storage, segregation and collection

Waste management includes collection, transport, processing, recycling and disposal of wastes. Of these the collection influences the rest of the operations thus significantly influences the overall waste management. Factors including environmental, economic, technical, legislational, and political issues play crucial role in shaping the waste management system and the collections. Collection systems are becoming complex than in the past due to the massive urbanization, which increases the importance of the waste collection in the modern society.

Street sweeping, waste collection and transport are the most publicly visible aspect of waste management to avoid nuisance and public health problems. In many low income countries, waste is often manually handled several times before final disposal at the point of arising, collection points and transfer stations. Waste collection activity accounts for about 80% of all the costs associated with waste management. Investment, operation and maintenance costs of vehicles, and salaries of collection workers comprise the vast majority of the costs of collection services.

The most effective waste management systems are those that effectively combine high service standards with high rates of materials recovery and recycling. This requires consideration of several issues such as the: (1) waste generation and characteristics (rates, density); (2) number of households and number of inhabitants per household; (3) existing storage facilities at households; (4) distance to collection/transfer point; (5) estimated number of required staff and vehicles; (6) existing facilities and infrastructure (storage facilities, roads, drains); (7) existing practices (waste pickers and waste recycling); (8) decision-making structure (area representatives, opinion leaders); (9) communication channels for information to households; (10) ability and willingness to contribute to waste management; (11) required equipments (containers, bins); (12) frequency of collection (storage capacity and odor emissions); (13) point of secondary collection (communal storage; house-to-house); and (14) integration of waste pickers.

2.1 Source segregation

The separation of recyclable material into individual components, either by the generator or by the collector at curbside, in different waste groups according to the specific treatment and disposal requirements is known as source segregation. Source separation of MSW becomes an integral component of planning and executing an integrated waste management system. Besides, its adoption is likely to increase in the coming years due to increasing preference for the recycling. It requires a high degree of homeowner involvement, and has high collection costs but low processing costs. Separately collected MSW fractions from the household can greatly assist recycling and resource recovery at a local level by providing a clean and reliable feedstock. Segregation should be carried out by the waste generator and it must take place at source. Source separated collection of wastes has several options, including selective collection of paper, glass, metal, plastics or the organic fraction as well as their multiple combinations. These options focus on the convenience of collecting recyclables separately in order to reduce recycling costs.

The quality of the segregation system will depend on the quality of the segregation concept and quality of the implementation of the concept. While considering to implement a source separate collection system, it should be kept in mind that the overall waste collection costs will be higher than that for mixed waste collection. Moreover, the success depends greatly on public participation which in turn will need sustained education and communication programs. Segregated waste should not be mixed during transport and storage. The sustenance of public participation requires the subsequent processing of wastes in segregated manners. Further issues associated with segregated collection of waste are: (1) special vehicles that might be needed for separate collection; (2) encouraging the establishment of formal or informal collection cooperatives; (3) public information and education schemes focused on encouraging householders, communities and businesses to segregate their wastes (either for the informal or formal sector).

2.2 Waste storage

Storage of waste at the source of waste generation till it is collected is the essential step of waste management. In residential areas where refuse is collected manually, either plastic bags or standard sized metal or plastic containers are typically required for waste storage. When automatic or semiautomatic collection systems are used, solid waste containers must be specifically designed to fit the truck-mounted loading mechanisms. The factors that must be considered in the on-site storage of solid wastes include the type of container to be used, container location, collection method and public health and aesthetic aspects. To a large extent, the types and capacities of containers depend on the characteristics of wastes to be stored, collection frequency and the space available for placement of the containers. The following good practices may be considered while choosing appropriate waste container systems:

(1) Containers shall be adequate in capacity, corrosion resistant and made of local, recyclable, or readily available materials;

(2) Containers that are easy to identify, due to either shape, colour or special markings are to be preferred;

(3) Containers shall be sturdy and/or easy to repair or replace;

(4) Containers shall prevent access by waste pickers and be appro-

priate to the terrain.

If community storage sites are being used, the storage points should be at intervals convenient to the generators and should be designed so that waste is deposited into containers at that site to facilitate secondary collection. With a view to ensuring that streets and public places are not littered with waste materials generated while on a move, litter bins may be provided on important streets, markets, public places, tourist spots, bus and railway stations, large commercial complexes, etc. at a distance ranging from 25 m to 250 m depending on the local condition.

2.3 Waste collection

Waste collection, often divided into "primary" and "secondary" services, is the process of picking up wastes from residences or collection points, loading into vehicles and transporting them to locations for processing, transfer or disposal. Primary collection is the means by which waste is collected from its source (dwellings and commercial premises) and transported to community storage, transfer points or even disposal sites. Secondary collection is the collection of solid waste for the second time, such as from community collection points, prior to its transport (often as part of a collection round by larger vehicles) to a transfer station, treatment facility or disposal site. Waste collection occupies the central position of an integrated waste management system. The way that waste materials are collected determines which waste management options can subsequently be used. Moreover, the selected collection method will significantly influence the quality of recovered materials.

Waste collection and transfer are often the most difficult components to design due to the involvement of many factors. To simplify the design of collection systems, the following factors must be considered and they also help to optimize an existing collection system: (1) define community goals and constrains; (2) characterize waste generation and service area; (3) determine public and private collection and transfer options; (4) determine system funding structure; (5) identify waste preparation and collection procedures; (6) identify collection equipment and crew size requirements; (7) evaluate transfer needs and options; (8) evaluate collection and transfer alternatives; (9) develop collection routes and schedules; (10) implement the collection system; (11) monitor system performance and adjust as necessary.

The management of collection is most difficult and complex in an urban environment because the generation of residential and commercial-industrial solid waste and recyclables takes place in every home, apartment building, commercial and industrial facility, as well as in the streets, parks, and vacant areas. As the patterns of waste generation becomes more diffuse and the total quantity of waste increases, the logistics of collection becomes more complex. A comparison of the different collection options are summarized in Table 3.

Table 3 Comparison of waste collection options

No.	Collection option	Operation	Comments
1	Door to door collection	Waste collector knocks on each door or rings doorbell and waits for waste to be brought out by resident	Convenient for residents. Little waste on streets. Residents must be available to hand waste over. Not suitable for apartment buildings because of the amount of walking required
2	Block collection	Collector sounds horn or rings bell and waits at specified locations for residents to bring waste to the collection vehicle	Economical. Less waste on streets. No permanent container or storage to cause complaints. If all family members are out when collector comes, waste must be left outside for collection, it may be scattered by wind, animals and waste pickers
3	Curbside collection	Waste is left outside property in a container and picked up by passing vehicle, or swept up and collected by sweeper	Convenient. No permanent public storage. Waste that is left out may be scattered by wind, animals, children or waste pickers
4	Back yard collection	Collection laborer enters property to remove waste	Very convenient for residents. No waste on streets. The most expensive system. Cultural beliefs, security considerations or architectural styles may prevent laborers from entering properties

Continued

No.	Collection option	Operation	Comments
5	Dumping at designated location (Drop-off centers)	Residents and other generators put their waste inside a container which is emptied or removed	Low capital costs. Loading the waste into trucks is slow and unhygienic. Waste is scattered around the collection point. Adjacent residents and shopkeepers protest about the smell and appearance
6	Shared container	Residents and other generators put their waste inside a container which is emptied or removed	Low operating costs. If containers are not maintained they quickly corrode or are damaged. Adjacent residents complain about the smell and appearance

Source: Jonathan W. C. Wong et al. (2016).

Key decisions of collection system design are the management and administration structure for the service standard of service to be provided, the agency undertaking collection (municipality, municipal enterprise, waste pickers, householders), the point of collection (from dwelling, curb, communal station), the types of vehicle to be used, the storage system and type/size/material of waste storage containers to be used and whether separation at source of recyclable materials is economically viable and must be allowed for and the frequency of collection.

2.3.1 Frequency of collection

Frequency of collection is a fundamental parameter of any waste collection system. Waste quantities arising, coupled with waste container capacities and climate conditions, customer expectations and cost will affect the frequency of collection. In hot and humid climates waste decomposes rapidly, posing an aesthetic and health risk. Denser communities may require more frequent waste collection services. Seasonal changes in waste generation will also require flexibility in the collection service. Some communities are accustomed to a collection seven days a week, whilst other collection agencies are striving for just once each week. If fly breeding is to be controlled, the waste should be collected twice a week in hot climates. Furthermore, the concentration of putrescible matter in wastes generated in economically developing countries is usually high.

2.3.2 Collection vehicles

Many technical factors have a direct bearing on the selection of a collection system and vehicle for any particular situation while the choices of vehicle and storage system are interrelated. The choice of waste collection vehicle is influenced by a range of factors including the waste generation rate, loading heights, stage of collection (i.e., primary or secondary), the method of solid waste collection adopted, the nature of the waste (density, volume, and composition), the topography of the city, the budget available for transport and the size (range) of the collection schedule, distance to disposal site, traffic restrictions and road conditions. The common types of primary collection vehicles in cities of developing countries are the: (1) handcart which is pushed by the operator as he walks along; (2) pedal tricycle with a tray or box in front of or behind the operator; (3) tractor-trailer units operated by a collector driver.

2.4 Waste separation

In case the waste is collected in mixed form, the recyclables with the general waste stream are recovered by manually or mechanically sorting at a suitable site such as material recycling facility (MRF) (Table 4). Such sorting by the municipality has the advantage of eliminating the dependence on the public and ensuring that the recycling does occur. The disadvantage, however, is that the value of the recyclable materials is reduced since being mixed in and compacted with other garbage can have adverse effects on the quality of the recyclable materials.

Table 4 Waste separation techniques

No.	Method of sorting	Brief description
1	Manual sorting	Manual sorting of waste is still very much a technique that is used in the world today. Furthermore, it is often a preliminary step in MRFs to remove the heavy objects which tend to affect the delicate instruments downstream

Continued

No.	Method of sorting	Brief description
2	Trommel separators/ Drum screens	The separation based on the size of the material is achieved using trommel and drum screens. Waste is fed into a large rotating drum which is perforated with holes of a certain size. Materials smaller than the diameter of the holes will be able to drop through, but larger particles will remain in the drum
3	Air classification	Separation of light and heavy fractions based on density using air as the media
4	Magnetic separator	Ferrous(magnetic) materials are separated from nonmagnetic materials
5	Eddy current separators	An "eddy current" occurs when a conductor is exposed to a changing magnetic field. Put simply, it is an electromagnetic way of dividing ferrous and nonferrous metals. This method is applied specifically to separate the metals
6	Induction sorting	Material is sent along a conveyor belt with a series of sensors underneath. These sensors locate different types of metal which are then separated by a system of fast air jets linked to the sensors
7	Near infrared (NIR) sensors	When materials are illuminated, they mostly reflect light in the near infrared wavelength spectrum. The NIR sensor can distinguish between different materials based on the way they reflect light
8	X-ray technology	X-rays can be used to distinguish between different types of waste based on their density (e.g. X-ray detection of PVC)

Source: Jonathan W. C. Wong et al. (2016).

Chapter 3 Status of municipal waste treatment and resource recovery

Municipal solid waste, commonly known as trash or garbage in the United States and as refuse or rubbish in Britain, is a waste type consisting of daily items that are discarded by the public. "Garbage" can also refer specifically to "food waste", as in a garbage disposal; however, the two are sometimes collected separately.

In the European Union, the semantic definition is "mixed municipal waste". Although the waste may originate from a number of sources that has nothing to do with a municipality, the traditional role of municipalities in collecting and managing these kinds of waste have produced the particular etymology "municipal".

3.1 Status of municipal waste treatment

3.1.1 Disposal

Today, the disposal of wastes by land filling or land spreading is the ultimate fate of all solid wastes, whether they are residential wastes collected and transported directly to a landfill site, residual materials from materials recovery facilities, residue from the combustion of solid waste, compost, or other substances from various solid waste processing facilities. A modern sanitary landfill is not a dump; it is an engineered facility used for disposing of solid wastes on land without creating nuisances or hazards to public health or safety, such as the problems of insects and the contamination of ground water.

3.1.2 Reusing

In the recent years environmental organizations, such as Freegle or Freecycle Network, have been gaining popularity for their online reuse networks. These networks provide a worldwide online registry of unwanted items that would otherwise be thrown away, for individuals and nonprofits to reuse or recycle. Therefore, this free Internet-based service reduces landfill pollution and promotes the gift economy.

3.1.3 Landfills

Landfills are created by land dumping. Land dumping method varies, most commonly it involves the mass dumping of waste into a designated area, usually a hole or side-hill. After the waste is dumped, it is then compacted by large machines. When the dumping cell is full, it is then "sealed" with a plastic sheet and covered in several feet of dirt. This is the primary method of dumping in the United States because of the low cost and abundance of unused land in North America. Landfills pose the threat of pollution, and can contaminate ground water. The signs of pollution are effectively masked by disposal companies and it is often hard to see any evidence. Usually landfills are surrounded by large walls or fences hiding the mounds of debris. Large amounts of chemical odor eliminating agent are sprayed in the air surrounding landfills to hide the evidence of the rotting waste inside the plant.

3.1.4 Energy generation

Municipal solid waste can be used to generate energy. Several technologies have been developed that make the processing of MSW for energy generation cleaner and more economical than ever before, including landfill gas capture, combustion, pyrolysis, gasification, and plasma arc gasification. While older waste incineration plants emitted a lot of pollutants, recent regulatory changes and new technologies have significantly reduced this concern. United States Environmental Protection Agency regulations in 1995 and 2000 under the Clean Air Act have succeeded in reducing emissions of dioxins from waste-to-energy facilities by more than 99% below 1990 levels, while mercury emissions have been reduced by over 90%. The EPA noted these improvements in 2003, citing waste-to-energy as a power source "with less environmental impact than almost any other source of electricity".

3.1.5　A case study in China

In China, along with urbanization, population growth and industrialization, the quantity of municipal solid waste generation has been increasing rapidly. The total MSW amount increased from 31.3 Mt in 1980 to 212 Mt in 2006, and the waste generation rate increased from 0.50 kg/(capita · d) in 1980 to 0.98 kg/(capita · d) in 2006. Currently, waste composition in China is dominated by a high organic and moisture content, since the concentration of kitchen waste in urban solid waste makes up the highest proportion (at approximately 60%) of the waste stream. The total amount of MSW collected and transported was 148 Mt in 2006, of which 91.4% was landfilled, 6.4% was incinerated and 2.2% was composted. The overall MSW treatment rate in China was approximately 62% in 2007. In 2007, there were 460 facilities, including 366 landfill sites, 17 composing plants, and 66 incineration plants.

3.2　Resource recovery

Recycling refers to the conversion of waste into valuable resources. For example, plastic is shredded and remodeled into new plastic products, and newspapers are recovered as the feedstock for cardboard or newspaper again. In an integrated waste management system, recycling is always listed near the top of a hierarchy, which was designed to achieve maximum economic, social and environmental returns. However, recycling alone is not enough to completely utilize the waste as a valuable resource.

Resource recovery refers to the process of acquiring energy or materials from wastes. It is a methodology constituted of reuse, recycling and waste-to-energy. Through resource recovery, the products are utilized optimally, virgin natural resources are saved, and the waste to be disposed is minimized. Table 5 summarizes some of the virgin resources saved by recovering the resources.

Table 5 Virgin resources saved by recovering the resource

Resource recovered	Virgin resources saved
Plastics	Recycling of plastics reduces the amount of crude oil for producing new plastics
Metals	Recycling of metal waste like aluminium drink cans will reduce the amount of bauxite (aluminium ore) required
Paper	Recycling of waste paper can reduce the amount of trees needed to produce new paper
Demolition waste	Recycling of demolition waste can substitute the consumption of natural sand and other building materials
Waste oil	Recycling of waste oil can substitute fuel oil or diesel as it can be converted into biodiesel

To achieve efficient resource recovery from waste, three consecutive steps are involved. The first step is the separation of resources from waste stream. There are two major separation approaches for solid waste: source separation and materials recovery facility. The second step is processing of the resources to materials or energy. For example, glass and ash can be processed, and used as construction materials. Finally, the third step is the utilization of the materials or energy by consumers or industries.

However, there are some issues concerning resource recovery practices. Firstly, lack of motivation to recycle. The lack of motivation is due to the disposal habits of consumers. Several solutions were devised to tackle this problem, and one of the solutions is to promote recycling by "pay-as-you-throw" schemes. Some companies like NTUC Fairprice have introduced a 10 cents rebate to customers who bring their own bags. Secondly, excessive costs and efforts are required to retrofit existing processes and facilities to suit resource recovery systems. Therefore, both technical and legislative measures should be addressed to promote a more robust resource recovery system. Table 6 gives an overview of the current challenges, processes and possible resource recovery solutions.

Table 6 Challenges, current processes, and solutions of resource recovery

Challenges	Current processes	Solutions
Disposal habits	(1) Small scale collection of waste by rag-and-bone man; (2) Excessive use of disposal materials, for example, plastic bags from supermarkets	(1) Disincentives and incentives to cultivate recycling habits; (2) Peer motivation and education
Technical and cost barriers	(1) Incineration of non-recycling waste; (2) Landfilling of waste which can not be incinerated	(1) Legislative measures to lower investment cost; (2) Research in new technologies to retrofit existing infrastructures

To sum up, waste is not waste, but a misplaced resource from our daily life and industrial manufacturing activities. With proper management and suitable technologies, waste can be recovered as reusable/new materials, energy, and other products with values. This resource recovery approach is a key component of modern solid waste management, and it provides both environmental and economic benefits.

Chapter 4 Analysis and determination of municipal solid waste

The importance of reliable information on the composition of municipal solid wastes is emphasized by the following facts:

(1) Potentials for recycling (e.g. composting, metals) or needs for treatment and disposal capacities (e.g. incinerators, landfills) can be identified only if information on the amount and composition of municipal solid waste is available;

(2) In order to design waste treatment processes properly, the materials to be treated have to be well characterized;

(3) Emissions to the environment from waste management practice can be predicted only if the inputs of waste treatment are known.

Because properties and composition of municipal solid wastes are changing constantly, it is necessary to analyze municipal solid waste periodically. This is especially necessary when many new consumer products are being introduced, which after a certain retention time in the anthroposphere will be discharged into the waste stream.

The parameters which are used to characterize waste materials can be divided into three groups: (1) materials (e.g. paper, glass, metals); (2) physical, chemical or biological parameters (e.g. density, water content, biodegradability); and (3) elemental concentrations (e.g. carbon, mercury).

To solve a particular problem of waste management, it is usually not necessary to characterize all parameters. It is often sufficient to analyze specific group of parameters. For example, for recycling studies, information on the content of different materials in municipal solid waste is needed. To predict emissions from thermal waste treatment, information on the elemental composition of municipal solid waste allows a first assessment of the environmental impacts. The most common approach to solid waste analysis is to collect a certain amount of municipal solid waste and take a number of samples, to screen, pulverize and analyze, and to calculate the waste composition from such procedures.

4.1 Physical properties of municipal solid waste

4.1.1 Specific weight (Density)

Specific weight is defined as the weight of a material per unit volume (e.g. kg/m^3, lb/ft^3), and usually it refers to uncompacted waste. It varies with geographic location, season of the year, and length of time in storage. The typical specific weight values are shown in Table 7.

Table 7 Typical specific weight values of municipal wastes

Component	Specific weight(Density)/(kg/m^3)	
	Range	Typical
Food wastes	130~480	290
Paper	40~130	89
Plastics	40~130	64
Yard waste	65~225	100
Glass	160~480	194
Tin cans	50~160	89
Aluminum cans	65~240	160
Condition	Density/(kg/m^3)	
Loose municipal solid waste with out processing or compaction	90~150	
In compaction truck	355~530	
Baled municipal solid waste	710~825	
Municipal solid waste in a compacted landfill(without cover)	440~740	

4.1.2 Moisture content

The moisture in a sample is expressed as percentage of the wet weight of the municipal solid waste material. The analysis procedure of moisture content is: (1) Weigh the aluminum dish; (2) Fill the dish with solid waste sample and re-weigh; (3) Dry solid waste + dish in an oven for at least 24 h at 105℃; (4) Remove the dish from the oven, and allow it to cool in a desiccator, and weigh; (5) Record the weight of the dry solid waste + dish; (6) Calculate the moisture content of the solid waste sample. Typical moisture contents of wastes are shown in Table 8.

Table 8 Typical moisture contents of wastes

Sources	Types of wastes	Moisture content/%	
		Range	Typical
Residential	Food wastes(mixed)	50~80	70
	Paper	4~10	6
	Plastics	1~4	2
	Yard wastes	30~80	60
	Glass	1~4	2
Commercial	Food wastes	50~80	70
	Rubbish(mixed)	10~25	15
Construction & Demolition	Mixed demolition combustibles	4~15	8
	Mixed construction combustibles	4~15	8
Industrial	Chemical sludge(wet)	75~99	80
	Sawdust	10~40	20
	Wood(mixed)	30~60	35
Agricultural	Mixed agricultural wastes	40~80	50
	Manure(wet)	75~96	94

4.1.3 Particle size and distribution

The size and distribution of the components of wastes are important for the recovery of materials, especially when mechanical means are used, such as trommel screens (Fig. 5) and magnetic separators. For example, ferrous items which are of a large size may be too heavy to be separated by a magnetic belt or drum system.

The size of waste components can be determined using the following

Fig. 5 Trommel screens used for recovery of materials from wastes

equations:

$$S_c = L$$
$$S_c = (L+W)/2$$
$$S_c = (L+W+H)/3$$

Where S_c——size of component, mm;

L——length, mm;

W——width, mm;

H——height, mm.

4.1.4 Field capacity

The total amount of moisture that can be retained in a waste sample subject to the downward pull of gravity, as shown in Fig. 6.

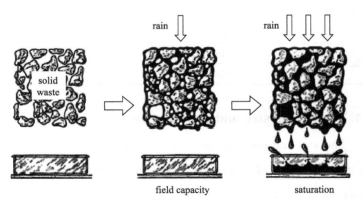

Fig. 6 Field capacity of materials from wastes

Field capacity is critically important in determining the formation of leachate in landfills. It varies with the degree of applied pressure and the state of decomposition of wastes, but typical values for uncompacted commingled wastes from residential and commercial sources are in the range of 50%~60%.

4.1.5 Permeability of compacted waste

The permeability (hydraulic conductivity) of compacted solid waste is an important physical property because it controls the movement of liquids and gases in a landfill. Permeability depends on: (1) pore size distribution; (2) surface area; (3) porosity, as shown in Fig. 7.

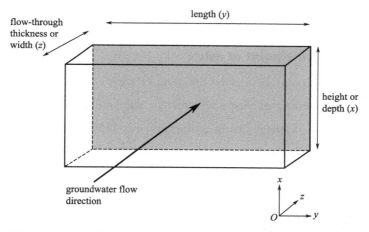

Fig. 7 Permeability that controls the movement of liquids or gases

4.2 Chemical properties of municipal solid waste

Chemical properties of municipal solid waste are very important in evaluating the alternative processing and recovery options, which can be divided into: (1) proximate analysis; (2) fusing point of ash; (3) ultimate analysis (major elements); (4) energy content.

4.2.1 Proximate analysis

Proximate analysis for the combustible components of municipal solid waste includes the following tests: (1) moisture (drying at 105℃ for 1 h); (2) volatile combustible matter (ignition at 950℃ in the absence of oxygen); (3) fixed carbon (combustible residue left after Step 2);

(4) ash (weight of residue after combustion in an open crucible). Typical proximate analysis values are shown in Table 9.

Table 9 Typical proximate analysis values of wastes   Unit: %

Type of waste	Moisture	Volatiles	Carbon	Ash
Mixed food	70.0	21.4	3.6	5.0
Mixed paper	10.2	75.9	8.4	5.4
Mixed plastics	0.2	95.8	2.0	2.0
Yard wastes	50.0	42.3	7.3	0.4
Glass	2.0	—	—	96~99
Residential municipal solid waste	21.0	52.0	7.0	20.0

4.2.2 Fusing point of ash

Fusing point of ash is the temperature at which the ash resulting from the burning of waste will form a solid (clinker) by fusion and agglomeration. Typical fusing temperature is 1100~1200℃. The formation of clinker from the burning of waste by fusion is shown in Fig. 8.

Fig. 8 Formation of clinker from the burning of waste by fusion

4.2.3 Ultimate analysis

Ultimate analysis involves the determination of the percent of C (carbon), H (hydrogen), O (oxygen), N (nitrogen), S (sulfur) and ash. The determination of halogens are often included in an ultimate analysis. The results are used to characterize the chemical composition of the organic matter in MSW. They are also used to define the proper mix of waste materials to achieve suitable C/N ratios for biological conversion processes. Chemical composition of typical municipal solid waste is shown in Fig. 9. Typical data on ultimate analysis of combustible materials found in solid waste is shown in Table 10, and typical data on elemental analysis of solid waste is shown in Table 11.

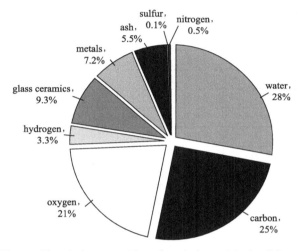

Fig. 9　Chemical composition of typical municipal solid waste

Table 10　Typical data on ultimate analysis of combustible materials in solid waste　　　　　Unit:%

Type of waste	Percent by weight(dry basis)					
	Carbon	Hydrogen	Oxygen	Nitrogen	Sulfur	Ash
Food and food products						
Fat	73.0	11.5	14.8	0.4	0.1	0.2
Food wastes(mixed)	48.0	6.4	37.6	2.6	0.4	5.0
Fruit wastes	48.5	6.2	39.5	1.4	0.2	4.2
Meat wastes	59.6	9.4	24.7	1.2	0.2	4.9
Paper products						
Cardboard	43.0	5.9	44.8	0.3	0.2	5.0
Magazines	32.9	5.0	38.6	0.1	0.1	23.3
Newsprint	49.1	6.1	43.0	<0.1	0.2	1.5
Paper(mixed)	43.4	5.8	44.3	0.3	0.2	6.0
Waxed cartons	59.2	9.3	30.1	0.1	0.1	1.2

Continued

Type of waste	Percent by weight(dry basis)					
	Carbon	Hydrogen	Oxygen	Nitrogen	Sulfur	Ash
Plastics						
Plastics(mixed)	60.0	7.2	22.8	—	—	10.0
Polyethylene	85.2	14.2	—	<0.1	<0.1	0.4
Polystyrene	87.1	8.4	4.0	0.2	—	0.3
Polyurethane	63.3	6.3	17.6	6.0	<0.1	4.3
Polyvinyl chloride	45.2	5.6	1.6	0.1	0.1	2.0
Textiles, rubber, leather						
Textiles	48.0	6.4	40.0	2.2	0.2	3.2
Rubber	69.7	8.7	—	—	1.6	20.0
Leather	60.0	8.0	11.6	10.0	0.4	10.0
Wood, trees, etc.						
Yard wastes	46.0	6.0	38.0	3.4	0.3	6.3
Wood(green timber)	50.1	6.4	42.3	0.1	0.1	1.0
Hardwood	49.6	6.1	43.2	0.1	<0.1	0.9
Wood(mixed)	49.5	6.0	42.7	0.2	<0.1	1.5
Wood chips(mixed)	48.1	5.8	45.5	0.1	<0.1	0.4
Glass, metals, etc.						
Glass and mineral	0.5	0.1	0.4	<0.1	—	98.9
Metals(mixed)	4.5	0.6	4.3	<0.1	—	90.5
Miscellaneous						
Office sweepings	24.3	3.0	4.0	0.5	0.2	68.0
Oils, paints	66.9	9.6	5.2	2.0	—	16.3
Refuse-derived fuel	44.7	6.2	38.4	0.7	<0.1	9.9

Table 11 Typical data on elemental analysis of solid waste (by weight)   Unit: %

Type	C	H	O	N	S	Ash
Mixed food	73.0	11.5	14.8	0.4	0.1	0.2
Mixed paper	43.3	5.8	44.3	0.3	0.2	6.0
Mixed plastic	60.0	7.2	22.8	—	—	10.0
Yard waste	46.0	6.0	38.0	3.4	0.3	6.3
Refuse-derived fuel	44.7	6.2	38.4	0.7	<0.1	9.9

4.2.4 Energy content

Energy content can be determined: (1) by using a full scale boiler as a calorimeter; (2) by using a laboratory bomb calorimeter (Fig. 10); (3) by calculation. Most of the data on the energy content of the organic components of MSW are based on the results of bomb calorimeter tests. Inert residue and energy content of residential municipal solid waste in Table 12, average composition and heating values for municipal solid waste in Table 13, energy content of municipal solid waste components in Fig. 11, and essential analysis of nutrients and other elements in Table 14.

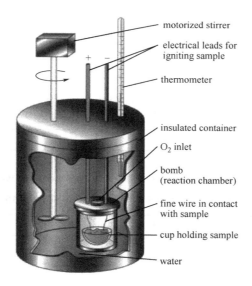

Fig. 10 Bomb calorimeter used for determination of energy content of solid waste

Table 12 Inert residue and energy content of residential municipal solid waste

Components	Inert residue percentage/%		Energy content/(kJ/kg)	
	Range	Typical	Range	Typical
Organics				
Food wastes	2~8	5	3489~6978	4652

Continued

Components	Inert residue percentage/%		Energy content/(kJ/kg)	
	Range	Typical	Range	Typical
Paper	4~8	6	11630~18608	16747
Cardboard	3~6	5	13956~17445	16282
Plastics	6~20	10	27912~37216	32564
Textiles	2~4	2.5	15119~18608	17445
Rubber	8~20	10	20934~27912	23260
Leather	8~20	10	15119~19771	17445
Yard wastes	2~6	4.5	2326~18608	6513
Wood	0.6~2	1.5	17445~19771	18608
Miscellaneous organics	—	—	—	—
Inorganics				
Glass	96~>99	98	116~233	140
Tin cans	96~>99	98	233~1163	698
Aluminum metal	90~>99	96	—	—
Other metal	94~>99	98	233~1163	698
Dirt, ashes, etc.	60~80	70	2326~11630	6978
Municipal solid waste(on average)	—	—	9304~13956	11630

Table 13 Average composition and heating values for municipal solid waste
(The average energy content of typical MSW is ~10,000 kJ/kg)

Waste component	Weight/%	Heating value/(MJ/kg)
Paper and paper products	37.8	17.7
Plastics	4.60	33.5
Rubber and leather	2.20	23.5
Textiles	3.30	32.5
Wood	3.00	20.0
Food wastes	14.2	15.1
Yard wastes	14.6	17.0
Glass and ceramics	9.00	0
Metals	8.20	0
Miscellaneous inorganics	3.10	0

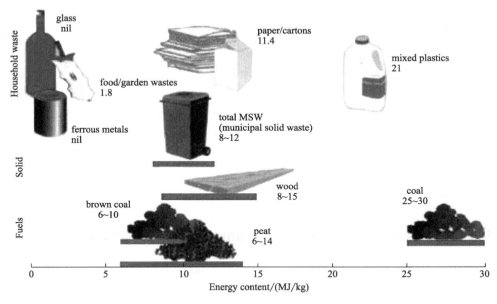

Fig. 11 Energy content of municipal solid waste components

Table 14 Essential analysis of the organic materials used as the feedstock for biological conversion processes (Important if organic fraction of municipal solid waste is to be used for production of compost or CH_4)

Constituent	Unit	Feed substrate (dry basis)			
		Newsprint	Office paper	Yard wastes	Food wastes
NH_4^+-N	mg/L	4	61	149	205
NO_3^--N	mg/L	4	218	490	4278
P	mg/L	44	295	3500	4900
PO_4^{3-}-P	mg/L	20	164	2210	3200
K	%	0.35	0.29	2.27	4.18
SO_4^{2-}-S	mg/L	159	324	882	855
Ca	%	0.01	0.10	0.42	0.43
Mg	%	0.02	0.04	0.21	0.16
Na	%	0.74	1.05	0.06	0.15
B	mg/L	14	28	88	17
Se	mg/L	—	—	<1	<1
Zn	mg/L	22	177	20	21

Continued

Constituent	Unit	Feed substrate(dry basis)			
		Newsprint	Office paper	Yard wastes	Food wastes
Mn	mg/L	49	15	56	20
Fe	mg/L	57	396	451	48
Cu	mg/L	12	14	7.7	6.9
Co	mg/L	—	—	5.0	3.0
Mo	mg/L	—	—	1.0	<1
Ni	mg/L	—	—	9.0	4.5
W	mg/L	—	—	4.0	3.3

4.3　Biological properties of municipal solid waste

The organic fraction of municipal solid waste (excluding plastics, rubber and leather) can be classified as: (1) water-soluble constituents—sugars, starches, amino acids and various organic acids; (2) hemicellulose—a product of 5-and 6-carbon sugars; (3) cellulose—a product of 6-carbon sugar (glucose); (4) fats, oils and waxes—esters of alcohols and long-chain fatty acids; (5) lignin—present in some paper products; (6) lignocellulose—combination of lignin and cellulose; (7) proteins—amino acid chains.

4.3.1　Biodegradability

The most important biological characteristic of the organic fraction of municipal solid waste is that almost all the organic components can be converted biologically to gases and relatively inert organic and inorganic solids.

The production of odors and the generation of flies are also related to the putrescible nature of the organic materials. These will be discussed when talking about landfill processes.

Volatile solids, determined by ignition at 550℃, is often used as a measure of the biodegradability of the organic fraction of municipal solid waste.

Some of the organic constituents of municipal solid waste are highly volatile but low in biodegradability (e. g. newsprint) due to lignin content.

The rate at which the various components can be degraded varies markedly. For practical purposes, the principal organic components in municipal solid waste are often classified as rapidly and slowly decomposable.

Biodegradable fraction of selected organic waste components can be calculated by:

$$BF = 0.83 - 0.028 \, LC$$

Where BF——biodegradable fraction expressed on a volatile solid basis;

  0.83——empirical constant;

  0.028——empirical constant;

  LC——lignin content of the volatile solid expressed as a percent of dry weight.

Data on the biodegradable fraction of selected organic waste components based on lignin content is shown in Table 15, and a calculation of biodegradable fraction of municipal solid waste is shown in Table 16.

Table 15 Data on the biodegradable fraction of selected organic waste components based on lignin content

Component	Volatile solids, percent of total solids/%	Lignin content, percent of volatile solid/%	Biodegradable fraction/%
Food wastes	7~15	0.4	0.82
Paper			
Newsprint	94.0	21.9	0.22
Office paper	96.4	0.4	0.82
cardboard	94.0	12.9	0.47
Yard wastes	50~90	4.1	0.72

Table 16 Calculation of biodegradable fraction of municipal solid waste

Component	Percent of municipal solid waste/%	Percent of biodegradable fraction in each component/%
Paper and cardboard	37.6	0.50
Glass	5.5	0

Continued

Component	Percent of municipal solid waste/%	Percent of biodegradable fraction in each component/%
Ferrous metals	5.7	0
Aluminum	1.3	0
Other nonferrous metals	0.6	0
Plastics	9.9	0
Rubber and leather	3.0	0
Textiles	3.8	0.5
Wood	5.3	0.7
Other materials	1.8	0.5
Food wastes	10.1	0.82
Yard trimmings	12.8	0.72
Miscellaneous inorganics	1.5	0.8
Total	98.9	—

4.3.2 Production of odors

Odors are developed when solid wastes are stored for long periods of time on-site between collection points, in transfer stations, and in landfills. It is more significant in warm climates. The formation of odors results from the anaerobic decomposition of the readily decomposable organic components found in municipal solid waste.

4.3.3 Physical transformations

The principal physical transformations that may occur in the operation of solid waste management systems include: (1) component separation; (2) mechanical volume reduction; (3) mechanical size reduction.

Physical transformations do not involve changes in phase (e.g., solid to gas), unlike chemical and biological transformations.

4.3.4 Chemical transformations

Chemical transformations of solid waste typically involve a change of phase (e.g., solid to liquid, solid to gas, etc.). To reduce the volume and/or to recover conversion products, the principal chemical processes used to transform municipal solid waste in thermal processes include: (1) combustion (chemical oxidation); (2) pyrolysis; (3) gasification.

4.3.5 Biological transformations

The biological transformations of the organic fraction of municipal solid waste may be used: (1) to reduce the volume and weight of the materials; (2) to produce compost; (3) to produce methane, and include: (1) aerobic composting; (2) low-solid anaerobic digestion; (3) high-solid anaerobic digestion (anaerobic composting).

4.3.6 Importance of transformations

Typically waste transformations are used: (1) to improve the efficiency of solid waste management systems; (2) to recover reusable and recyclable materials; (3) to recover conversion products and energy.

The organic fraction of municipal solid waste can be converted to usable products and ultimately to energy in a number of ways including: (1) combustion to produce steam and electricity; (2) pyrolysis to produce a synthetic gas, liquid or solid fuel, and solids; (3) gasification to produce a synthetic fuel; (4) biological conversion to produce compost; (5) biodigestion to generate methane and to produce a stabilized organic humus. Transformation processes in municipal solid waste management are shown in Table 17.

Table 17 Transformation processes in municipal solid waste management

	Processes	Methods	Principal conversion products
Physical transformation	separation	manual and/or mechanical	individual component found in commingled MSW
	volume reduction	Force or pressure	original waste reduced in volume
	size reduction	Shredding, grinding, or milling	altered in form and reduced in size
Chemical transformation	combustion	thermal oxidation	CO_2, SO_2, oxidation products, ash
	pyrolysis	destructive distillation	a variety of gases, tars and/or oils
	gasification	starved air combustion	gases and inerts

Continued

	Processes	Methods	Principal conversion products
Biological transformation	aerobic compost	aerobic biological conversion	compost
	anaerobic digestion	anaerobic biological conversion	methane, CO_2, trace gases, humus
	anaerobic composting (in landfills)	anaerobic biological conversion	methane, CO_2, digested waste

Chapter 5 Landfill treatment for municipal solid waste

Landfill is the most popularly used method of waste disposal in many countries. This process focuses on burying the waste in the land. Landfills have the potential to cause a number of issues. Infrastructure disruption, such as damage to access roads by heavy vehicles, may occur. Pollution of local roads and water courses from wheels on vehicles when they leave the landfill can be significant and can be mitigated by wheel washing systems. Pollution of the local environment, such as contamination of groundwater or aquifers or soil may occur, as well. Extensive efforts are made to capture and treat leachate from landfills before it reaches groundwater aquifers, but engineered liners always have a lifespan, though it may be 100 years or longer.

Methane is naturally generated by decaying organic wastes in a landfill. It is a potent greenhouse gas, and can itself be a danger because it is flammable and potentially explosive. In properly managed landfills, gas is collected and utilized. This could range from simple flare to landfill gas utilization. As well, poorly run landfills may become nuisances because of carriers such as rats and flies which can cause infectious diseases.

5.1 What is landfill?

A landfill (also known as a tip, dump, rubbish dump, garbage dump or dumping ground, and historically as a midden) is a site for the disposal of waste materials by burial and the oldest form of waste treatment (although the burial part is modern; historically, refuse was just left in piles or thrown into pits). Historically, landfills have been the most common method of organized waste disposal and remain so in many places around the world.

Some landfills are also used for waste management purposes, such as the temporary storage, consolidation and transfer, or processing of waste materials (sorting, treatment, or recycling). Unless they are stabilized, these areas may experience severe shaking or soil liquefaction of the ground during a large earthquake.

5.2 Types of landfills

The level of environmental protection depends on the type of waste accepted at the facility. The types of landfills are shown in Table 18.

Table 18 Types of landfills

Types of landfills	Brief description
Municipal solid waste landfills	Landfills that accept household waste as well as other wastes
Managed landfills	Landfills composed mainly of cleanfill materials, but also construction and demolition waste with light contaminants
Construction and demolition landfills	Landfills where construction and demolition materials such as wood products, asphalt, plasterboard, insulation materials and others are disposed to land
Cleanfills	Landfills where cleanfill material is disposed to land. Cleanfill material is material that when buried will have no adverse effect on people or the environment. It includes virgin natural materials such as clay, soil and rock, and other inert materials such as concrete or brick that are free of: (1) combustible, putrescible, degradable or leachable components; (2) hazardous substances; (3) products or materials derived from hazardous waste treatment, hazardous waste stabilisation or hazardous waste disposal practices; (4) materials that may present a risk to human or animal health such as medical and veterinary waste, asbestos or radioactive substances; (5) liquid waste
Industrial landfills	Landfills that accept specified industrial wastes. In most cases industrial waste landfills are monofills associated with a specific industry or facility

5.3 Operations of landfills

Typically, operators of well-run landfills for non-hazardous waste meet predefined specifications by applying techniques to: (1) confine waste to as small an area as possible; (2) compact waste to reduce volume; (3) cover waste (usually daily) with layers of soil or other types of materials such as woodchips and fine particles.

During landfill operations a scale or weighbridge may weigh waste collection vehicles on arrival and personnel may inspect loads for wastes that do not accord with the landfill's waste-acceptance criteria. Afterward, the waste collection vehicles use the existing road network on their way to the tipping face or working front, where they unload their contents. After loads are deposited, bulldozers or compactors can spread and compact the waste on the working face. Before leaving the landfill boundaries, the waste collection vehicles may pass through a wheel-cleaning facility. If necessary, they return to the weighbridge for re-weighing without their loads. The weighing process can assemble statistics on the daily incoming waste tonnage, which databases can retain for record keeping. In addition to trucks, some landfills may have equipment to handle railroad containers. The use of "rail-haul" permits landfills to be located at more remote sites, without the problems associated with many truck trips.

Typically, in the working face, the compacted waste is covered with soil or alternative materials daily. Alternative waste-cover materials include chipped wood or other "green waste", several spray-on foam products, chemically "fixed" bio-solids, and temporary blankets. Blankets can be lifted into place at night and then removed the following day prior to waste placement. The space that is occupied daily by the compacted waste and the cover material is called a daily cell. Waste compaction is critical to extending the life of the landfill. Factors such as waste compressibility, waste-layer thickness and the number of passes of the compactor over the waste affect the waste density.

5.4　Advantages of landfills

Landfills are often the most cost-efficient way to dispose of waste, especially in countries like the United States with large open spaces. While resource recovery and incineration both require extensive investments in infrastructure, and material recovery also requires extensive manpower to maintain, landfills have lower fixed or ongoing costs, allowing them to compete favorably. In addition, landfill gas can be upgraded to natural gas—landfill gas utilization—which is a potential revenue stream. Another advantage is having a specific location for disposal that can be monitored, where waste can be processed to remove all recyclable materials before tipping.

5.5　Social and environmental impact of landfills

Landfills have the potential to cause a number of issues. Infrastructure disruption, such as damage to access roads by heavy vehicles, may occur. Pollution of local roads and water courses from wheels on vehicles when they leave the landfill can be significant and can be mitigated by wheel washing systems. Pollution of the local environment, such as contamination of groundwater aquifers or soil may occur, as well.

5.5.1　Leachate

Extensive efforts are made to capture and treat leachate from landfills before it reaches groundwater aquifers, but engineered liners always have a lifespan, though it may be 100 years or longer. Eventually, every landfill liner will leak, allowing the leachate to contaminate the groundwater. Installation of composite liners with flexible membrane and soil barrier is enforced by the EPA to ensure that leachate is withheld.

5.5.2　Dangerous gases

Rotting food and other decaying organic waste allows methane and carbon dioxide to seep out of the ground and up into the air. Methane is a potent greenhouse gas, and can itself be a danger because it is flammable and potentially explosive. In properly managed landfills, gas is collected and utilized. This could range from simple flare to landfill gas utilization. Carbon dioxide is the most widely produced greenhouse gas. It traps heat in the atmosphere, contributing to climate change.

5.5.3 Infections

Poorly run landfills may become nuisances because of carriers such as rats and flies which can cause infectious diseases. The occurrence of such carriers can be mitigated by the use of daily cover.

Other potential issues include wildlife disruption, dust, odor, noise pollution, and reduced local property values.

5.6 Landfill gas

Landfill gas is a complex mixture of different gases created by the action of microorganisms within a landfill. Landfill gas is approximately $40\% \sim 60\%$ methane, with the remainder being mostly carbon dioxide. Trace amounts of other volatile organic compounds comprise the remainder ($<1\%$). These trace gases include a large array of species, mainly simple hydrocarbons.

Gases are produced in landfills due to the anaerobic digestion by microbes. In a properly managed landfill this gas is collected and used. Its usage ranges from simple flare to the landfill gas utilization and generation of electricity. Landfill gas monitoring alerts workers to the presence of a build-up of gases to a harmful level. In some countries, landfill gas recovery is extensive; in the United States, for example, more than 850 landfills have active landfill gas recovery systems.

Landfill gases have an influence on climate change. The major components are CO_2 and methane, both of which are greenhouse gas. In terms of global warming potential (GWP), methane is over 25 times more detrimental to the atmosphere than carbon dioxide. Landfills are the third largest source of human-made methane in the US.

5.6.1 Landfill gas production

Landfill gases are the result of three processes: (1) evaporation of volatile organic compounds (e.g., solvents); (2) chemical reactions between waste components; (3) microbial actions, especially methanogenesis.

The first two depend strongly on the nature of the waste. The dominant process in most landfills is the third process whereby anaerobic bacteria decompose organic waste to produce biogas, which consists of methane and carbon dioxide together with traces of other compounds. Despite the heterogeneity of waste, the evolution of gases follows well defined kinetic pattern. Formation of methane and CO_2 commences about six months after depositing the landfill materials. The evolution of gas reaches a maximum in about 20 years, and then declines over the course of decades.

5.6.2 Monitoring of landfill gas

Since the gases produced by landfills are both valuable and sometimes hazardous, monitoring techniques have been developed. Flame ionization detectors can be used to measure methane levels as well as total VOC levels. Surface monitoring and sub-surface monitoring as well as monitoring of the ambient air is carried out. In the U.S., under the Clean Air Act of 1996, it is required that many large landfills install gas collection and control systems, which means that at the very least the facilities must collect and flare the gas.

U.S. federal regulations under Subtitle D of RCRA formed in October 1979 regulate the siting, design, construction, operation, monitoring, and closure of MSW landfills. Subtitle D now requires controls on the migration of methane in landfill gas. Monitoring requirements must be met at landfills during their operation, and for an additional 30 years afterward. The landfills affected by Subtitle D of RCRA are required to control gas by establishing a way to check for methane emissions periodically and therefore prevent off-site migration. Landfill owners and operators must make sure that the concentration of methane gas does not exceed 25% of the LEL for methane in the facility structures and the LEL for methane at the facility boundaries.

5.6.3 Use of landfill gas

U. S. Environmental Protection Agency data shows that more than 950 municipal solid waste landfills are operating in the United States as of 2013. Decomposing wastes in these landfills produce landfill gas, which is a mixture of about half methane and half carbon dioxide. These landfills are the third largest source of human-made methane emissions in the United States.

The gases produced within a landfill can be collected and used in various ways. The landfill gas can be utilized directly on site by a boiler or any type of combustion system, providing heat (Fig. 12). Electricity can also be generated on site through the use of microturbines, steam turbines, or fuel cells. The landfill gas can also be sold off site and sent into

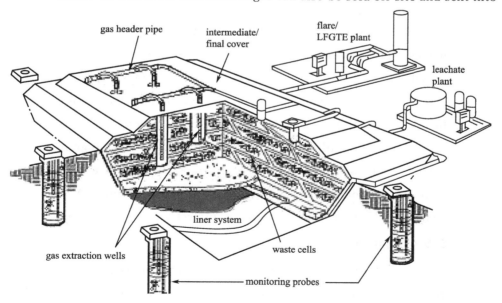

Fig. 12　Landfill gas collection system

natural gas pipelines. This approach requires the gas to be processed into pipeline quality, e. g. , by removing various contaminants and components. The efficiency of gas collection at landfills directly impacts the amount of energy that can be recovered—closed landfills (those no longer accepting waste) collect gas more efficiently than open landfills (those still accepting waste). A comparison of collection efficiency at closed and open landfills found about 17% difference between the two.

Landfill gas can also be used to evaporate leachate, another byproduct of the landfill process, displacing another fuel that was previously used.

In the U. S. , the number of landfill gas projects increased from 399 in 2005 to 594 in 2012 according to the Environmental Protection Agency. These projects are popular because they control energy costs and reduce greenhouse gas emissions. These projects collect the methane gas and treat it, so it can be used for electricity or upgraded to pipeline-grade gas. Methane gas has twenty-one times the global warming potential of carbon dioxide. For example, in the U. S. , Waste Management uses landfill gas as an energy source at 110 landfill gas-to-energy facilities. This energy production offsets almost 2Mt of coal per year, creating energy equivalent to that needed by four hundred thousand homes. These projects also reduce greenhouse gas emissions into the atmosphere.

The U. S. Environmental Protection Agency, which estimates that hundreds of landfills could support gas to energy projects, has also established the Landfill Methane Outreach Program. This program was developed to reduce methane emissions from landfills in a cost-effective manner by encouraging the development of environmentally and economically beneficial landfill gas-to-energy projects.

5.6.4 Safety

Landfill gas emissions can lead to environmental, hygiene and security problems in the landfill. Several accidents have occurred, for example at Loscoe, England in 1986, where migrating landfill gas accumulated and partially destroyed a property. An accident causing two deaths occurred from an explosion in a house adjacent to Skellingsted Landfill in Denmark in 1991. Due to the risk presented by landfill gas, there is a clear need to monitor gas produced by landfills. In addition to the risk of fire and explosion, gas migration in the subsurface can result in contact of landfill gas with groundwater. This, in turn, can result in contamination of groundwater by organic compounds present in nearly all landfill gas.

Although usually evolving only trace amounts, landfills do release some aromatics and chlorocarbons.

Landfill gas migration, due to pressure differentials and diffusion, can occur. This can create an explosion hazard if the gas reaches sufficiently high concentrations in adjacent buildings.

5.7 Landfill leachate production, collection and management

5.7.1 Leachate production

In modern landfills, the waste is actually encapsulated and sealed off to prevent migration of pollutants and pathogens. The key parts of landfill design are: (1) the cover system to minimize air emissions and infiltration of precipitation and (2) the bottom liner and leachate collection system to prevent leachate from getting outside the landfill boundaries or down into groundwater. It is important to keep the landfill as dry as possible to reduce the amount of leachate. This is done by collecting surface runoff with ditches and catch basins, and the collected runoff is sent to collection ponds, treated for suspended soil particles and possible contaminants, and then pumped off-site.

It is unavoidable for a landfill to generate leachate due to (1) rainwater soaking into the landfill and (2) the moisture within the trash itself. Leachate seeps downward to the base of the landfill by gravity, carrying particulate materials in suspension along with the dissolved constituents. Usually, the Hydrologic Evaluation of Landfill Performance (HELP) model is used to estimate leachate generation. The HELP program was developed to conduct water balance analyses of landfills, cover systems, solid waste disposal and containment facilities. The model facilities allow rapid estimation of the amounts of runoff, evapotranspiration, drainage, leachate collection, and liner leakage that may be expected as a result of the operation of a wide variety of landfill designs. The model, applicable to open, partially closed, and fully closed sites, is a tool for designers to "assist in the comparison of design alternatives as judged by their water balances".

It should be pointed out that the model: (1) should not be used for sizing leachate treatment facilities; (2) should be used with particular caution under different weather conditions; (3) is very sensitive to the allowance of dry periods and evapotranspiration, and accurately estimated evapotranspiration of the site can be a major limitation; and (4) omits some conditions in open landfills and seldom considers landfills with a final cover. It is important to realize that climate changes in recent years have caused many problems for rainfall intensity and return periods, which calls for new evaluation on how to conduct water balance analyses of landfills. For example, the HELP version 3 was examined in a German climate in an extensive validation study, and a German adaptation was developed. However, for arid or semi-arid regions, or locations with frequent extreme weather conditions, the HELP model may have limitations, weak points, and errors, and thus, requires further research.

5.7.2 Leachate collection

Landfills usually have a multiple layer system to collect the highly contaminated leachate. The system consists of a filter zone (a geotextile), a primary leachate collection zone (with a leachate collection piping system), a primary barrier liner (flexible membrane liner), a secondary leachate collection zone (with another leachate collection piping system), a secondary barrier liner, and a compacted clay liner.

In the past, tremendous efforts have been made on developing materials that are suitable for use in order to satisfy drainage and filter requirements. Geotextiles (also called filter fabrics) and geosynthetic drainage nets (geonets) have been developed and used to replace soil drainage layers to save the space in the filter zone. Long-term creep and clogging are major design considerations. Currently, it is unknown exactly what criteria should be used in assessing the potential for the clogging of geonets. Wastewater quality criteria for evaluation of drip irrigation clogging hazards is recommended by others.

Leachate is collected and conveyed through a network of pipes and sumps. The leachate collection pipe is sized to handle the anticipated flow with Manning equation:

$$Q = 1.486 A R^{2/3} S^{0.5}/n$$

Where, Q is flow rate; A is the cross-sectional area; R is hydraulic radius; S is slope, and n is Manning roughness coefficient. Assuming a circular plastic pipe (with $n=0.01$) running full, the pipe diameter (D) can be calculated as:

$$D = 0.327 (Q/S^{0.5})^{0.375}$$

The leachate is directed to a separate leachate collection pond for contaminants detection and then follow-up treatment. The treatment may occur on-site or off-site. The options for treatment include recirculating the leachate back to the landfill, treating for sanitary sewer discharge and for local surface water discharge.

Considerable research has been conducted on materials that can be used as the liner. Liners are broadly classified as: (1) soil liners, e.g., compacted clay liners, synthetic clay liners, or soil (soil sand) mixed with bentonite; (2) geomembranes; and (3) geosynthetic clay liners where soil (mixed with bentonite) is sandwiched between two sheets of fabric. The current dilemma for landfill liners are that "once the unit is closed, the bottom layer of the landfill will deteriorate over time and, consequently, will not prevent the transport of leachate out of the unit". It is commonly recognized that even the most advanced landfill liner systems would eventually fail to prevent groundwater pollution by municipal solid waste leachate. Therefore, one important direction is to develop a more sophisticated leachate detection system located between the two composite liners. When leachate is detected in that layer, action will be taken to prevent leachate from leaking out the landfill. Moreover, it would require the availability of adequate post-closure funding and responsibility to maintain the cover, operate the leachate collection system and leak detection system, and the ability to effect repairs in all systems, for as long as the wastes in the landfill can generate leachate when it is contacted by water.

5.7.3 Leachate management

Resources Conservation and Recovery Act (passed in 1976 by the US Congress) regulations restrict leachate head on the liner to 30 cm for Subtitle D landfills. Leachate in excess of 30 cm should be removed from the landfill. Leachate can be removed by using gravity flow or by pumping. Leachate collected from the landfill may be stored on site to be treated later, or transported for treatment and disposal off-site. Surface impoundments and tanks are the typical leachate storage methods. The most economical option is to transport leachate to an off-site facility for treatment and disposal. This allows the owners/operators to focus on their primary goal of managing the landfill while the leachate is handled by experts on wastewater treatment. From the owner/operator's perspective, this option also eliminates the burden associated with the permitting, testing, monitoring and reporting requirements.

Leachate treatment is challenging mainly because of the irregular production rates and variable compositions. A summary of commonly used treatment methods is presented in Table 19. More than one method is often required to achieve the intended goal. Leachate treatment practiced at the Al Turi Landfill in New York is an example for use of multiple methods. Polymer coagulation, flocculation, sedimentation, anaerobic biological treatment, aerobic biological treatment, and filtration were used in the process.

Table 19 Summary of leachate treatment options

Treatment option	Removal objective	Comments
Biological—Best used on "young" leachate		
Activated sludge	BOD/COD	Flexible, shock resistant, proven, minimum SRT increases with increasing organic strength, $>90\%$ BOD removal possible
Aerated lagoons	BOD/COD	Good application to small flows, $>90\%$ BOD removal possible
Anaerobic	BOD/COD	Aerobic polishing necessary to achieve high-quality effluent
Powdered activated carbon/activated sludge	BOD/COD	$>95\%$ COD removal, $>99\%$ BOD removal
Physical/Chemical—Useful as polishing step or for treatment of "old" leachate		
Coagulation/Precipitation	Heavy metals	High removal of Fe, Zn; moderate removal of Cr, Cu, Mn; low removal of Cd, Pb, Ni
Chemical oxidation	COD	Raw leachate treatment requires high chemical dosages, better used as polishing step
Ion exchange	COD	$10\% \sim 70\%$ COD removal, slight metal removal
Adsorption	BOD/COD	$30\% \sim 70\%$ COD removal after biological or chemical treatment
Reverse osmosis	TDS	$90\% \sim 96\%$ TDS removal

Notes: 1. COD—Chemical oxygen demand;

2. BOD—Biological oxygen demand;

3. TDS—Total dissolved solids.

Source: Vesilind et al. (2002); King and Mureebe (1992).

Chapter 6 Thermal treatment for municipal solid waste

The disposal of solid waste has become increasingly intricate and costly, because of the decrease in space available for landfills and the growing concern about the living environment. Thermal treatment has the merit to reduce the volume of waste and to render it innocuous, by destruction of pathogens and toxic chemicals. The composition and structure of solid waste are destroyed thoroughly at high temperature, with the principles of thermal stability.

The thermal technologies (shown in Fig. 13) are essential components of the integrated solid waste management systems, and are also already operating successfully as confirmed by several studies. They are characterized by higher temperatures and conversion rates than most others, i. e. the biochemical and physicochemical processes, and they can dispose different types of solid waste, in particular the unsorted residual municipal solid waste. Their main benefits are: (1) significant reduction of waste volume (typically, about 80%~90%) and mass (typically, about 70%~80%); (2) short processing time, from few minutes (fluidized bed units) to more than one hour (mechanical grate or stoker units) or even a large part of a day (some pyrolysis units); (3) destruction of organic contaminants, such as halogenated hydrocarbons (dioxins and furans), polycyclic aromatic hydrocarbons (PAHs), phenols, cyanides; (4) concentration and immobilization of inorganic contaminants, such as heavy metals, which may be usefully and safely utilized or else disposed; (5) thorough disinfection, which is a good method for disposal of pathogenic waste and organic-polluted waste; (6) utilization of ferrous and non-ferrous metals from fly and bottom ash; (7) avoided environmental burdens as evidenced through life cycle assessment studies (waste-to-energy was shown to have less environmental impact than almost any other source of electricity); (8) utilization of the renewable energy of solid waste, such as electricity and process heat, and the cost of the operation can be offset to some extent by energy sales. Thermal treatment technology includes three main thermal conversion processes: combustion, gasification and pyrolysis, which are compared schematically in Table 20. Other thermal treatments include roasting (disposal of solid waste under the melting point), dewatering and drying, thermal decomposition, and sintering.

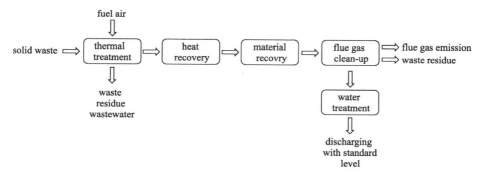

Fig. 13 Basic flow chart of thermal treatment processes

Table 20 Main characteristics of the chemical processes for thermal treatment of solid waste

	Combustion	Pyrolysis	Gasification
Aim of the process	To maximize waste conversion to high temperature flue gases, mainly CO_2 and H_2O	To maximize thermal decomposition of solid waste to gases and condensed phases	To maximize waste conversion to high-heating-value fuel gases, mainly CO, H_2, and CH_4
Operating conditions (Reaction environment)	Oxidizing (oxidant amount larger than that required by stoichiometric combustion)	Total absence of any oxidant	Reducing (oxidant amount lower than that required by stoichiometric combustion)

Continued

	Combustion	Pyrolysis	Gasification
Reactant gas	Air	None	Air, pure oxygen, oxygen-enriched air, steam
Temperature	850~1200℃	500~800℃	550~900℃ (in air gasification), 1000~1600℃
Pressure	Atmospheric	Slight over-pressure	Atmospheric
Types of reactors	Grate furnace, fluidized bed incinerator and rotary kiln incinerator	Batch, semi-batch and continuous reactors; fixed bed, fluidized bed and screw kiln reactors	Fixed bed, fluidized bed, rotary kiln bed gasifiers and other gasifiers
Process output (Produced gases)	CO_2, H_2O	CO, H_2, CH_4 and other hydrocarbons	CO, H_2, CO_2, H_2O, CH_4
Pollutants	SO_2, NO_x, HCl, PCDD/PCDF, particulate	H_2S, HCl, NH_3, HCN, tar, particulate	H_2S, HCl, COS, NH_3, HCN, tar, alkali, particulate
Gas cleaning	Treated in air pollution control units to meet the emission limits and then sent to the stack	It is possible to clean the syngas to meet the standards of chemical production processes or those of efficient energy conversion devices	It is possible to clean the syngas to meet the standards of chemical production processes or those of efficient energy conversion devices
Ash	Bottom ash can be treated to recover ferrous (iron, steel) and non-ferrous metals (such as aluminium, copper and zinc) and inert materials (to be utilized as a sustainable building material). Air pollution control residues (APCRs) are generally treated and disposed as industrial waste	Often having un-negligible carbon content. Treated and disposed as industrial special waste	As for combustion process. Bottom ash are often produced as vitreous slag that can be utilized as backfilling material for road construction

Source: Arena (2012); Arena and Mastellone (2009).

6.1 Incineration

Incineration with energy recovery is one of several waste-to-energy technologies such as gasification, pyrolysis and anaerobic digestion. While incineration and gasification are similar in principle, the energy product from incineration is high-temperature heat whereas combustible gas is often the main energy product from gasification. Incineration and gasification may also be implemented without energy and materials recovery.

Incinerators reduce the solid mass of the original waste by 80% ~ 85% and the volume (already compressed somewhat in garbage trucks) by 95% ~ 96%, depending on composition and recovery degree of materials such as metals from the ash for recycling. This means that while incineration does not completely replace landfilling, it significantly reduces the necessary volume for disposal. Garbage trucks often reduce the volume of waste in a built-in compressor before delivery to the incinerator. Alternatively, at landfills, the volume of the uncompressed garbage can be reduced by approximately 70% by using a stationary steel compressor, albeit with a significant energy cost. In many countries, simpler waste compaction is a common practice for compaction at landfills.

Incineration has particularly strong benefits for the treatment of certain types of wastes such as clinical wastes and certain hazardous wastes in niche areas where pathogens and toxins can be destroyed by high temperatures. Examples include chemical multi-product plants with diverse toxic or very toxic wastewater streams, which cannot be routed to a conventional wastewater treatment plant.

Waste combustion is particularly popular in countries such as Japan where land is a scarce resource. Denmark and Sweden have been leaders by using the energy generated from incineration for more than a century, supporting district heating schemes in localised combined heat and power facilities. In 2005, waste incineration produced 4.8% of the electricity consumption and 13.7% of the total domestic heat consumption in Denmark. A number of other European countries rely heavily on incineration for handling municipal waste, in particular Luxembourg, the Netherlands, Germany, and France.

6.1.1 History of incineration

The first UK incinerators for waste disposal were built in Nottingham by Manlove, Alliott & Co. Ltd. in 1874, and the design was patented by Albert Fryer. They were originally known as destructors.

The first US incinerator was built in 1885 on Governors Island in New York, NY. The first facility in the Czech Republic was built in 1905 in Brno.

6.1.2 Incineration technologies

An incinerator is a furnace for burning wastes. Modern incinerators include pollution mitigation equipment such as flue gas cleaning system. There are various types of incinerator plant designs: moving grate, fixed grate, rotary kiln, and fluidised bed.

(1) Burn pile

The burn pile is one of the simplest and earliest forms of waste disposal, essentially consisting of a mound of combustible materials piled on open ground and set on fire, as shown in Fig. 14.

Fig. 14 A typical small burn pile in a garden

Burn piles can and have spread uncontrolled fires, for example if wind blows burning material off the pile into surrounding combustible grasses or onto buildings. As interior structures of the pile are consumed, the pile can shift and collapse, spreading the burn area. Even in a situation of no wind, small lightweight ignited embers can lift off the pile via convection, and waft through the air into grasses or onto buildings, igniting them. Burn piles often do not result in full combustion of waste and therefore produce particulate pollution.

(2) Burn barrel

The burn barrel is a somewhat more controlled form of private waste incineration, containing the burning material inside a metal barrel, with a metal grating over the exhaust. The barrel prevents the spread of burning material in windy conditions, and as the combustibles are reduced they can only settle down into the barrel. The exhaust grating helps to prevent the spread of burning embers. Typically 55-US-gallon (210 L) steel drums are used as burn barrels, with air vent holes cut or drilled around the base for air intake. Over time, the very high heat of incineration causes the metal to oxidize and rust, and eventually the barrel itself is consumed by the heat and must be replaced.

Private burning of dry cellulosic/paper products is generally clean-burning, producing no visible smoke, but plastics in household waste can cause private burning to create a public nuisance, generating acrid odors and fumes that make eyes burn and tear. Most urban communities ban burn barrels, and certain rural communities may have prohibitions on open burning, especially those home to many residents not familiar with this common rural practice.

As of 2006 in the United States, private rural household or farm waste incineration of small quantities was typically permitted so long as it is not a nuisance to others, and does not pose a risk of fire such as in dry conditions, and the fire does not produce dense, noxious smoke. A handful of states, such as New York, Minnesota, and Wisconsin, have laws or regulations either banning or strictly regulating open burning due to health and nuisance effects. People intended to burn waste may be required to contact a state agency in advance to check current fire risk and conditions, and to alert officials of the controlled fire that will occur.

(3) Moving grate

The typical incineration plant for municipal solid waste is a moving grate incinerator. The moving grate enables the movement of waste through the combustion chamber to be optimized to allow a more efficient and complete combustion. A single moving grate boiler can handle up to 35 t (39 short tons) of waste per hour, and can operate 8,000 h per year with only one scheduled stop for inspection and maintenance of about one month's duration. Moving grate incinerators are sometimes referred to as Municipal Solid Waste Incinerators. A control room of a typical moving grate incinerator overseeing two boiler lines is shown in Fig. 15. Municipal solid waste in the furnace of a moving grate incinerator is shown in Fig. 16, and the holes in the grate elements supplying the primary combustion air are visible.

Fig. 15 Control room of a typical moving grate incinerator overseeing two boiler lines

Fig. 16 Municipal solid waste in the furnace of a moving grate incinerator capable of handling 15 metric tons (17 short tons) of waste per hour

The waste is introduced by a waste crane through the "throat" at one end of the grate, from where it moves down over the descending grate to the ash pit in the other end. Here the ash is removed through a water lock.

Part of the combustion air (primary combustion air) is supplied through the grate from below. This air flow also has the purpose of cooling the grate itself. Cooling is important for the mechanical strength of the grate, and many moving grates are also water-cooled internally.

Secondary combustion air is supplied into the boiler at high speed through nozzles over the grate. It facilitates complete combustion of the flue gases by introducing turbulence for better mixing and by ensuring a surplus of oxygen. In multiple/stepped hearth incinerators, the secondary combustion air is introduced in a separate chamber downstream the primary combustion chamber.

According to the European Waste Incineration Directive, incineration plants must be designed to ensure that the flue gases reach a temperature of at least 850℃ (1,560°F) for 2 s in order to ensure proper breakdown of toxic organic substances. In order to comply with this at all times, it is required to install backup auxiliary burners (often fueled by oil), which are fired into the boiler in case the heating value of the waste becomes too low to reach this temperature alone.

The flue gases are then cooled in the superheaters, where the heat is transferred to steam, heating the steam to typically 400℃ (752°F) at a pressure of 4MPa (580 psi) for the electricity generation in the turbine. At this point, the flue gas has a temperature of around 200℃ (392°F), and is passed to the flue gas cleaning system.

In Scandinavia, scheduled maintenance is always performed during summer, When the demand for district heating is low. Often, incineration plants consist of several separate "boiler lines" (boilers and flue gas treatment plants), so that waste can continue to be received at one boiler line while the others are undergoing maintenance, repairing, or upgrading.

(4) Fixed grate

The older and simpler kind of incinerator was a brick-lined cell with a fixed metal grate over a lower ash pit, with one opening on the top or side for loading and another opening on the side for removing incombustible solids called clinkers. Many small incinerators formerly found in apartment houses have now been replaced by waste compactors.

(5) Rotary kiln

The rotary kiln incinerator is used by municipalities and by large industrial plants. This design of incinerator has 2 chambers: a primary chamber and a secondary chamber. The primary chamber in a rotary kiln incinerator consists of an inclined refractory-lined cylindrical tube. The inner refractory lining serves as sacrificial layer to protect the kiln structure. This refractory layer needs to be replaced from time to time. Movement of the cylinder on its axis facilitates movement of waste. In the primary chamber, there is conversion of solid fraction to gases, through volatilization, destructive distillation and partial combustion reactions. The secondary chamber is necessary to complete gas phase combustion reactions.

The clinkers spill out at the end of the cylinder. A tall flue-gas stack, fan, or steam jet supplies the needed ventilation. Ash drops through the grate, while many particles are carried along with the hot gases. The particles and any combustible gases may be combusted in an "afterburner".

(6) Fluidized bed

A strong airflow is forced through a sandbed. The air seeps through the sand until a point is reached where the sand particles separate to let the air through and mixing and churning occurs, thus a fluidized bed is created, and fuel and waste can now be introduced. The sand with the pretreated waste and/or fuel is kept suspended in pumped air currents and takes on a fluid-like character. The bed is thereby violently mixed and agitated keeping small inert particles and air in a fluid-like state. This allows all of the mass of waste, fuel and sand to be fully circulated through the furnace.

(7) Specialized incineration

Sawdust incinerators in furniture factories need much attention as they have to handle resin powder and many flammable substances. Controlled combustion and burn back preventive systems are essential as dust when suspended resembles the fire phenomenon of any liquid petroleum gas.

6.1.3 Use of heat

The heat produced by an incinerator can be used to generate steam which may then be used to drive a turbine in order to produce electricity. The typical amount of net energy that can be produced by incinerating per tonne of municipal waste is about 2/3 MW · h of electricity and 2 MW · h of district heating. Thus, incinerating about 600 t (660 short tons) of waste per day will produce about 400 MW · h of electrical energy per day (17 MW of electrical power continuously for 24 h) and 1200 MW · h of district heating energy each day.

6.1.4 Pollution

Incineration has a number of outputs such as the ash and the emission of flue gas to the atmosphere. Before the flue gas cleaning system, if installed, the flue gases may contain particulate matter, heavy metals, dioxins, furans, sulfur dioxide, and hydrochloric acid. If plants have inadequate flue gas cleaning, these outputs may add a significant pollution component to stack emissions.

In a study from 1997, Delaware Solid Waste Authority found that, for the same amount of produced energy, incineration plants emitted fewer particles, hydrocarbons and less SO_2, HCl, CO and NO_x than coal-fired power plants, but more than natural gas-fired power plants. According to Germany Federal Ministry for the Environment, waste incinerators reduce the amount of some atmospheric pollutants by substituting power produced by coal-fired plants with power from waste-fired plants.

6.1.5 Gaseous emissions

(1) Dioxin and furans

The most publicized concerns from environmentalists about the incineration of municipal solid wastes involve the fear that it produces significant amounts of dioxin and furan emissions. Dioxins and furans are considered by many to be serious health hazards. The EPA announced in 2012 that the safe limit for human oral consumption is 0.7 pg/(kg·d) Toxic Equivalence, which works out to 17×10^{-9} g/a for a 150 lbs person (1lb=0.45kg).

In 2005, Germany Federal Ministry for the Environment, where there were 66 incinerators at that time, estimated that "... whereas in 1990 one third of all dioxin emissions in Germany came from incineration plants, for the year 2000 the figure was less than 1%. Chimneys and tiled stoves in private households alone discharge approximately 20 times more dioxin into the environment than incineration plants."

According to the United States Environmental Protection Agency, the combustion percentages of the total dioxin and furan inventory from all known and estimated sources in the U.S. (not only incineration) for each type of incineration are as follows: 35.1% backyard barrels; 26.6% medical waste; 6.3% municipal wastewater treatment sludge; 5.9% municipal waste combustion; 2.9% industrial wood combustion. Thus, the controlled combustion of waste accounted for 41.7% of the total dioxin inventory.

In 1987, before the governmental regulations required the use of emission controls, there was a total of 8,905.1 g (314.12 oz) Toxic Equivalence of dioxin emissions from US municipal waste combustors. Today, the total emissions from the plants are 83.8 g (2.96 oz) Toxic Equivalence annually, with a reduction of 99%.

Backyard barrel burning of household and garden wastes, still allowed in some rural areas, generates 580 g (20 oz) of dioxins annually. Studies conducted by the USEPA demonstrated that the emissions from just one family using a burn barrel produced more emissions than an incineration plant disposing of 200 t (220 short tons) of waste per day by 1997 and five times that by 2007 due to increased chemicals in household trash and decreased emissions by municipal incinerators using advanced technology.

However, the same researchers found that their original estimates for the burn barrel were high, and that the incineration plant used for comparison represented a theoretical "clean" plant rather than any existing facility. Their later studies found that burn barrels produced a median of 24.95 ng Toxic Equivalence per lb garbage burned, so that a family burning 5 lbs of trash per day, or 1825 lbs per year, produces a total of 0.0455 mg Toxic Equivalence per year, and that the equivalent number of burn barrels for the 83.8 g (2.96 oz) of the 251 municipal waste combustors inventoried by the EPA in the U.S. in 2000, is 1,841,700, or on average, 7,337 family burn barrels per municipal waste incinerator.

Most of the improvement in U.S. dioxin emissions has been for large-scale municipal waste incinerators. As of the year 2000, although small-scale incinerators (those with a daily capacity of less than 250 t) processed only 9% of the total waste combusted, they produced 83% of the dioxins and furans emitted by municipal waste combustion.

(2) Dioxin cracking methods and limitations

The breakdown of dioxin requires exposure of the molecular ring to a sufficiently high temperature so as to trigger the thermal breakdown of the strong molecular bonds holding it together. Small pieces of fly ash may be somewhat thick, and too brief an exposure to high temperature may only degrade dioxin on the surface of the ash. For a large volume air chamber, too brief an exposure may also result in only some of the exhaust gases reaching the full breakdown temperature. For this reason there is also a time factor in the temperature exposure to ensure heating is completely achieved through the thickness of the fly ash and the volume of waste gases.

There are trade-offs between increasing either the temperature or exposure time. Generally where the molecular breakdown temperature is higher, the exposure time for heating can be shorter, but excessively high temperatures can also cause wear and damage to other parts of the incineration equipment. Likewise the breakdown temperature can be lowered to some degree but then the exhaust gases would require a greater lingering period of perhaps several minutes, which would require larger/longer treatment chambers that take up a great deal of treatment plant space.

A side effect of breaking the strong molecular bonds of dioxin is the potential for breaking the bonds of nitrogen gas (N_2) and oxygen gas (O_2) in the supply air. As the exhaust flow cools, these highly reactive detached atoms spontaneously reform bonds into reactive oxides such as NO_x in the flue gas, which can result in smog and acid rain if they were released directly into the local environment. These reactive oxides must be further neutralized with selective catalytic reduction or selective non-catalytic reduction.

(3) Dioxin cracking in practice

The temperatures needed to break down dioxin are typically not reached when burning plastics outdoors in a burn barrel or garbage pit, causing high dioxin emissions as mentioned above. While plastic does usually burn in an open-air fire, the dioxins remain after combustion and either float off into the atmosphere, or may remain in the ash where it can be leached down into groundwater when rain falls on the ash pile. Fortunately, dioxin and furan compounds bond very strongly to solid surfaces and are not dissolved by water, so leaching processes are limited to the first few millimeters below the ash pile. The gas-phase dioxins can be substantially destroyed using catalysts, some of which can be present as part of the fabric filter bag structure.

Modern municipal incinerator designs include a high-temperature zone, where the flue gas is sustained at a temperature above 850 ℃ (1,560°F) for at least 2 seconds before it is cooled down. The municipal incinerators are equipped with auxiliary heaters to ensure this at all times. They are often fueled by oil or natural gas, and are normally only active for a very small fraction of the time. Further, most modern incinerators utilize fabric filters (often with Teflon membranes to enhance collection of sub-micron particles) which can capture dioxins present in or on solid particles.

For very small municipal incinerators, the required temperature for thermal breakdown of dioxin may be reached using a high-temperature electric heating element, plus a selective catalytic reduction stage.

Although dioxins and furans may be destroyed by combustion, their reformation by a process known as "de novo synthesis" as the emission gases cool is a probable source of the dioxins measured in emission stack tests from plants with high combustion temperatures and long residence times.

(4) CO_2

As for other complete combustion processes, nearly all of the carbon content in the waste is emitted as CO_2 to the atmosphere. MSW contains approximately the same mass fraction of carbon as CO_2 itself (27%), so incineration of 1 ton of municipal solid waste produces approximately 1 ton of CO_2.

If landfilled, 1 ton of municipal solid waste would produce approximately 62 m^3 (2,200 ft^3) of methane via the anaerobic decomposition of the biodegradable part of the waste. Since the global warming potential of methane is 34 and the weight of 62 m^3 of methane at 25℃ is 40.7 kg, equivalent to 1.38 tons of CO_2, which is more than the 1 ton of CO_2 which would have been produced by incineration. In some countries, large amounts of landfill gas are collected. Still the global warming potential of the landfill gas emitted to atmosphere is significant. In the US, it was estimated that the global warming potential of the emitted landfill gas in 1999 was approximately 32% higher than the amount of CO_2 that would have been emitted by incineration. Since this study, the global warming potential estimate for methane has been increased from 21 to 35, which alone would increase this estimate to almost the triple GWP effect compared to incineration of the same waste.

In addition, nearly all the biodegradable waste has biological source. This material has been formed by plants using atmospheric CO_2 typically within the last growing season. If these plants are regrown, the CO_2 emitted from their combustion will be taken out of the atmosphere once more.

Such considerations are the main reason why several countries administrate incineration of biodegradable waste as renewable energy. The rest-mainly plastics and other oil and gas derived products-is generally treated as non-renewable resource.

Different results for the CO_2 footprint of incineration can be obtained with different assumptions. Local conditions (such as limited local district heating demand, no fossil-fuel-generated electricity to replace, or high levels of aluminium in the waste stream) can decrease the CO_2 benefits of incineration. The methodology and other assumptions may also influence the results significantly. For example, the methane emissions from landfills occurring at a later date may be neglected or given less weight, or biodegradable waste may not be considered as CO_2 neutral. A study by *Eunomia Research and Consulting* in 2008 on potential waste treatment technologies in London demonstrated that by applying several of these (according to the authors) unusual assumptions, the average existing incineration plants performed poorly for CO_2 balance compared to the theoretical potential of other emerging waste treatment technologies.

(5) Other emissions

Other gaseous emissions in the flue gas from incinerator furnaces include nitrogen oxides, sulfur dioxide, hydrochloric acid, heavy metals, and fine particles. Among the heavy metals, mercury is a major concern due to its toxicity and high volatility, as essentially all the mercury in the municipal waste stream may exit in emissions if not removed by emission controls.

The steam content in the flue may produce visible fume from the stack, which can be perceived as a visual pollution. It may be avoided by decreasing the steam content by flue-gas condensation and reheating, or by increasing the flue gas outlet temperature well above its dew point. Flue-gas condensation allows the latent heat of evaporation of the water to be recovered, subsequently increasing the thermal efficiency of the plant.

6.1.6 Flue-gas cleaning

The quantity of pollutants in the flue gas from incineration plants may or may not be reduced by several processes, depending on the plant.

Particulate is collected by particle filtration, most often electrostatic precipitators and/or baghouse filters. The latter are generally very efficient for collecting fine particles. In an investigation by the Ministry of the Environment of Denmark in 2006, the average particulate emissions from 16 Danish incinerators were below 2.02 g/GJ (grams per energy content of the incinerated waste). Detailed measurements of fine particles with sizes below 2.5 μm ($PM_{2.5}$) were performed on three of the incinerators: One incinerator equipped with an electrostatic precipitators for particle filtration emitted 5.3 g/GJ fine particles, while the other two incinerators equipped with baghouse filters emitted 0.002 g/GJ and 0.013 g/GJ $PM_{2.5}$. For ultra-fine particles ($PM_{1.0}$), the numbers were 4.889 g/GJ $PM_{1.0}$ from the plant equipped with electrostatic precipitators, while emissions of 0.000 g/GJ and 0.008 g/GJ $PM_{1.0}$ were measured from the plants equipped with baghouse filters.

Acid gas scrubbers are used to remove hydrochloric acid, nitric acid, hydrofluoric acid, mercury, lead and other heavy metals. The removal efficiency will depend on the specific equipment, the chemical composition of the waste, the design of the plant, the chemistry of reagents, and the ability of engineers to optimize these conditions, which may conflict for different pollutants. For example, mercury removal by wet scrubbers is considered coincidental and may be less than 50%. Alkali scrubbers remove sulfur dioxide, forming gypsum by reaction with lime.

Waste water from scrubbers must subsequently pass through a sewage treatment plant.

Sulfur dioxide may also be removed by dry desulfurization by injecting limestone slurry into the flue gas before the particle filtration.

NO_x is either reduced by catalytic reduction with ammonia in a catalytic converter (selective catalytic reduction) or by a high-temperature reaction with ammonia in the furnace (selective non-catalytic reduction). Urea may substitute for ammonia as the reductant but it must be supplied earlier in the process so that it can be hydrolyzed into ammonia. Substitution of urea can reduce costs and potential hazards associated with storage of anhydrous ammonia.

Heavy metals are often adsorbed on injected active carbon powder, which is collected by particle filtration.

6.1.7 Solid outputs

Incineration produces fly ash and bottom ash just as is the case when coal is combusted (as shown in Fig. 17). The total amount of ash produced by municipal solid waste incineration ranges from 4% to 10% by volume and 15%~20% by weight of the original quantity of waste, and the fly ash amounts to about 10%~20% of the total ash. The fly ash, by far, constitutes more of a potential health hazard than the bottom ash because the fly ash often contains high concentrations of heavy metals such as lead, cadmium, copper and zinc as well as small amounts of dioxins and furans. The bottom ash seldom contains significant levels of heavy metals. In tests over the past decade, no ash from an incineration plant in the USA has ever been identified as a hazardous waste. Currently, although some historical samples tested by the incinerator operators' group would meet the being ecotoxic criteria, at present the EA stated that "we have agreed" to regard the incinerator bottom ash as "non-hazardous" until the testing programme is complete.

Fig. 17 Operation of an incinerator aboard an aircraft carrier

6.1.8 Other pollution issues

Odor pollution can be a problem with old-style incinerators, but odors and dust are extremely well controlled in newer incineration plants. They receive and store the waste in an enclosed area with a negative pressure with the airflow being routed through the boiler which prevents unpleasant odors from escaping into the atmosphere. However, not all plants are implemented this way, resulting in inconveniences in the locality.

An issue that affects community relationships is the increased road traffic of waste collection vehicles to transport municipal waste to the incinerator. For this reason, most incinerators are located in industrial areas. This problem can be avoided to an extent through the transport of waste by rail from transfer stations.

6.2 Pyrolysis

Pyrolysis is a thermal decomposition of materials at elevated temperatures in an inert atmosphere such as a vacuum or nitrogen gas. It involves the change of chemical composition and is irreversible. The word is coined from the Greek-derived elements pyro for "fire" and lysis for "separating". A simplified depiction of pyrolysis chemistry is shown in Fig. 18.

Pyrolysis is most commonly applied to the treatment of organic materials. It is one of the processes involved in charring wood, starting at 200~300℃ (390~570°F). In general, pyrolysis of organic substances produces volatile products and leaves a solid residue enriched in carbon, i.e. char. Extreme pyrolysis, which leaves most of the carbon as the residue, is called carbonization.

The process is used extensively in the chemical industry, for example, to produce ethylene, many forms of carbon, and other chemicals from petroleum, coal, and even wood, to produce coke from coal. Aspirational applications of pyrolysis would convert biomass into

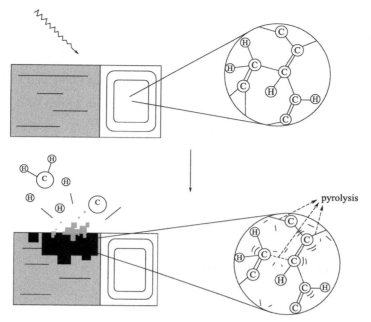

Fig. 18 Simplified depiction of pyrolysis chemistry

syngas and biochar, waste plastics back into usable oil, or waste into safely disposable substances.

6.2.1 Process terminology and mechanism

Certain uses of pyrolysis are called dry distillation, destructive distillation, or cracking. The processes involve thermal depolymerization, i.e. the breaking of chemical bonds in macromolecules to give smaller fragments. The phenomenon involves exceeding the ceiling temperature of polymerization.

Pyrolysis differs from other processes like combustion and hydrolysis in that it usually does not involve the addition of other reagents such as oxygen (O_2, in combustion) or water (in hydrolysis). In practice, it is often impractical to achieve a completely O_2—or water—free conditions, especially as pyrolysis is often conducted on complex mixtures. The term has also been applied to describing the decomposition of organic materials in the presence of superheated water or steam (hydrous pyrolysis), for example, in the steam cracking of oil. Pyrolysis has been assumed to take place during catagenesis, the conversion of buried organic matter to fossil fuels. In vacuum pyrolysis, organic material is heated in a vacuum to decrease its boiling point and avoid adverse chemical reactions, which is called flash vacuum pyrolysis, and this approach is used in organic synthesis.

Pyrolysis can be analyzed by pyrolysis gas chromatography mass spectrometry (Py-GC-MS). In that technique, the volatile products from pyrolysis are separated by gas chromatography, and identified by MS. The technique is applied to many aspects of pyrolysis technology. Pyrolysis is also used in carbon-14 dating.

6.2.2 Occurrence and uses

(1) Ethylene

Pyrolysis is used to produce ethylene, the chemical compound produced on the largest scale industrially (more than 110 Mt/a in 2005). In this process, hydrocarbons from petroleum are heated to around 600℃ in the presence of steam, i.e. steam cracking. The resulting ethylene is used to make antifreeze (ethylene glycol), PVC (via vinyl chloride), and many polymers, such as polyethylene and polystyrene.

(2) Coke, carbon, charcoals, and chars

Since carbon and carbon-rich materials have desirable properties but are nonvolatile, even at high temperatures, consequently, pyrolysis is used to produce many kinds of carbon, as a fuel, as a reagent in steelmaking (coke), and as a structural material.

High temperature pyrolysis is used on an industrial scale to convert coal into coke for metallurgy, especially steelmaking. Typical organic products obtained by pyrolysis of coal (X = CH, N) is shown in Fig. 19. Volatile products are often useful, including benzene and pyridine. Coke can also be produced from the solid residue left from petroleum refining.

The coke-making or "coking" process consists of heating the materi-

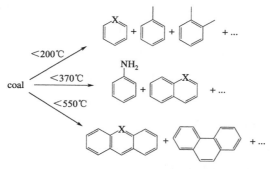

Fig. 19　Typical organic products obtained by pyrolysis of coal (X=CH, N)

al in "coking ovens" to very high temperatures (up to 900℃ or 1,700°F) so that those molecules are broken down into lighter volatile substances, which will leave the vessel, and a porous but hard residue that is mostly carbon and inorganic ash. The amount of volatiles varies with the source material, but is typically 25%～30% of it by weight.

The original vascular structure of the wood and the pores formed by escaping gases combine to produce a light and porous material. By starting with a dense wood-like material, such as nutshells or peach stones, a form of charcoal can be obtained with particularly fine pores (and hence a much larger pore surface area), called activated carbon, which is used as an adsorbent for a wide range of chemical substances.

Biochar is the residue of incomplete organic pyrolysis, e. g., from cooking fires. It is a key component of the terra preta soils associated with ancient indigenous communities of the Amazon basin. Terra preta is much sought by local farmers for its superior fertility compared to the natural red soil of the region. Efforts are underway to recreate these soils through biochar, i. e. the solid residue of pyrolysis of various materials which are mostly organic waste.

Carbon fibers are filaments of carbon that can be used to make very strong yarns and textiles. Carbon fiber items are often produced by spinning and weaving the desired item from fibers of a suitable polymer, and then pyrolyzing the material at a high temperature (1,500~3,000℃ or 2,730~5,430°F). The first carbon fibers were made from rayon, but polyacrylonitrile has become the most common starting material. For their first workable electric lamps, Joseph Wilson Swan and Thomas Edison used carbon filaments made from pyrolysis of cotton yarns and bamboo splinters, respectively.

Pyrolysis is the reaction used to coat a preformed substrate with a layer of pyrolytic carbon. This is typically done in a fluidized bed reactor heated to 1,000~2,000℃ or 1,830~3,630°F. Pyrolytic carbon coatings are used in many applications, including artificial heart valves.

(3) Biofuels

Pyrolysis is the basis of several methods for producing fuel from biomass, i.e. lignocellulosic biomass. Crops studied as biomass feedstock for pyrolysis include native North American prairie grasses such as switchgrass and bred versions of other grasses such as Miscanthus giganteus. Other sources of organic materials as feedstock for pyrolysis include greenwaste, sawdust, waste wood, nut shells, straw, cotton trash and rice hulls. Animal wastes including poultry litter, dairy manure, and potentially other manures are also under evaluation. Some industrial byproducts are also suitable feedstock including paper sludge and distillers' grains.

Synthetic diesel fuel by pyrolysis of organic materials is not yet economically competitive. Higher efficiency is sometimes achieved by flash pyrolysis, in which finely divided feedstock is quickly heated to a temperature between 350℃ and 500℃ (660°F and 930°F) for less than 2 seconds.

The low quality of oils produced through pyrolysis can be improved by physical and chemical processes, which might drive up production costs, but may make sense economically as circumstances change.

There is also the possibility of integrating with other processes such as mechanical biological treatment and anaerobic digestion. Fast pyrolysis is also investigated for biomass conversions. Fuel bio-oil can also be produced by hydrous pyrolysis.

6.2.3 Semiconductors

The process of metal-organic vapor phase epitaxy entails pyrolysis of volatile organometallic compounds to give semiconductors, hard coatings, and other applicable materials. The illustration of the metal-organic vapor phase epitaxy process is shown in Fig. 20. The reactions entail thermal degradation of precursors, with deposition of the inorganic component and release of the hydrocarbons as gaseous waste. Since it is an atom-by-atom deposition, these atoms organize themselves into crystals to form the bulk semiconductor. Silicon chips are produced by the pyrolysis of silane:

$$SiH_4 \longrightarrow Si + 2H_2$$

Gallium arsenide, another semiconductor, forms upon co-pyrolysis of trimethylgallium and arsine.

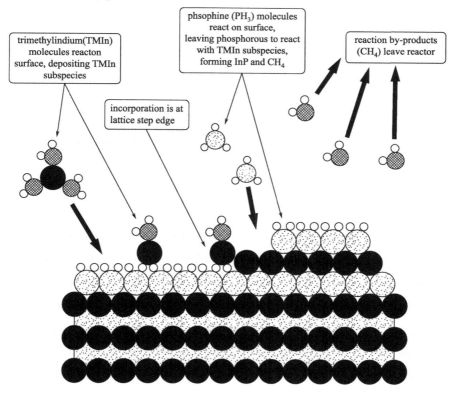

Fig. 20 Illustration of the metal-organic vapor phase epitaxy process, which entails pyrolysis of volatile

(1) Recycling

Pyrolysis can also be used to treat plastic waste. The main advantage is the reduction in volume of the waste. In principle, pyrolysis will regenerate the monomers (precursors) of the polymers that are treated, but in practice the process is neither clean nor economically competitive source of monomers.

Tire recycling is a well developed technology. Other products from car tire pyrolysis include steel wires, carbon black and bitumen. The field faces legislative, economic, and marketing obstacles. Oil derived from tire rubber pyrolysis contains high sulfur content, which gives it high potential as a pollutant and should be desulfurized.

(2) Thermal cleaning

Pyrolysis is also used for thermal cleaning, an industrial application to remove organic substances such as polymers, plastics and coatings from parts, products or production components like extruder screws, spinnerets and static mixers. During the thermal cleaning process, at temperatures between 600°F to 1000°F (310℃ to 540℃), organic material is converted by pyrolysis and oxidation into volatile organic compounds, hydrocarbons and carbonized gas, while inorganic elements remain.

Several types of thermal cleaning systems use pyrolysis: (1) Molten Salt Baths belong to the oldest thermal cleaning systems; cleaning with a molten salt bath is very fast but implies the risk of dangerous splatters, or other potential hazards connected with the use of salt baths, like explosions or highly toxic hydrogen cyanide gas. (2) Fluidized Bed Systems use sand or aluminium oxide as heating medium; these systems also clean very fast but the medium does not melt or boil, nor emit any vapor or odor; the cleaning process takes 1 to 2 hours. (3) Vacuum Ovens use pyrolysis in a vacuum avoiding uncontrolled combustion inside the cleaning chamber; the cleaning process takes 8 to 30 hours. (4) Burn-Off Ovens, also known as Heat-Cleaning Ovens, are gas-fired and used in the painting, coatings, electric motors and plastics industries for removing organics from heavy and large metal parts.

6.2.4 Fine chemical synthesis

Pyrolysis is used in the production of chemical compounds, mainly, but not only, in the research laboratory.

The field of boron-hydride clusters started with the study of the pyrolysis of diborane (B_2H_6) at ca. 200℃. Products include pentaborane and decaborane clusters. These pyrolyses involve not only cracking (to give H_2), but also recondensation.

The synthesis of nanoparticles, zirconia and oxides utilizing an ultrasonic nozzle in a process is called ultrasonic spray pyrolysis.

6.2.5 History

Pyrolysis has been used for turning wood into charcoal since ancient times. Many important chemical substances, such as phosphorus and sulfuric acid, were first obtained by this process. In their embalming process, the ancient Egyptians used methanol, which they obtained from the pyrolysis of wood. The dry distillation of wood remained the major source of methanol by the early 20th century.

6.3 Gasification

Gasification is a process that converts organic (or fossil fuel)-based carbonaceous materials into carbon monoxide, hydrogen and carbon dioxide. This is achieved by reacting the material at high temperatures ($>700℃$), without combustion, with a controlled amount of oxygen and/or steam. The resulting gas mixture is called syngas (from synthesis gas) or producer gas and is itself a fuel. The power derived from gasification and combustion of the resultant gas is considered to be a source of renewable energy if the gasified compounds are obtained from biomass.

The advantage of gasification is that using the syngas (synthesis gas H_2/CO) is potentially more efficient than direct combustion of the original fuel because it can be combusted at higher temperatures or even in fuel cells, so that the thermodynamic upper limit to the efficiency defined by Carnot's rule is higher or (in the case of fuel cells) not applicable. Syngas may be burned directly in gas engines, used to produce methanol and hydrogen, or converted into synthetic fuel via the Fischer-Tropsch synthesis. Gasification can also begin with material such as biodegradable waste which would otherwise have been disposed of. In addition, the high-temperature process refines out corrosive ash elements such as chloride and potassium, allowing clean gas production from otherwise problematic fuels. Gasification of fossil fuels is currently widely used on industrial scales to generate electricity.

6.3.1 History

The process of producing energy by gasification has been in use for more than 180 years. In the early time, coal and peat were used to power these plants. Initially developed to produce town gas for lighting and cooking in the 1800s, it was replaced by electricity and natural gas. It was also used in blast furnaces but the bigger role was played in the production of synthetic chemicals where it has been in use since the 1920s.

During both world wars, especially the World War II, the need for fuel produced by gasification reemerged due to the shortage of petroleum. Wood gas generators, called Gasogene or Gazogène, were used to power motor vehicles in Europe. Adler Diplomat 3 with gas generator (1941) is shown in Fig. 21. By 1945 there were trucks, buses and agricultural machines that were powered by gasification. It is estimated that there were close to 9,000,000 vehicles running on producer gas all over the world.

6.3.2 Chemical reactions

In a gasifier, the carbonaceous material undergoes several different processes:

(1) The dehydration or drying process occurs at around 100℃.

Fig. 21　Adler Diplomat 3 with gas generator (1941)

Typically the resulting steam is mixed into the gas flow and may be involved with subsequent chemical reactions, notably the water-gas reaction if the temperature is sufficiently high.

(2) The pyrolysis (or devolatilization) process (shown in Fig. 22) occurs at around 200~300℃. Volatiles are released and char is produced, resulting in up to 70% weight loss for coal. The process is dependent on the properties of the carbonaceous material and determines the structure and composition of the char, which will then undergo gasification reactions.

(3) The combustion process (shown in Fig. 23) occurs as the volatile products and some of the char react with oxygen to primarily form carbon dioxide and small amounts of carbon monoxide, which provides heat for the subsequent gasification reactions. Letting C represent a carbon-containing organic compound, the basic reaction here is $C + O_2 \longrightarrow CO_2$.

(4) The gasification process (shown in Fig. 23) occurs as the char reacts with steam to produce carbon monoxide and hydrogen, via the reaction $C + H_2O \longrightarrow H_2 + CO$.

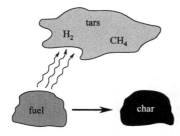
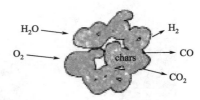

Fig. 22　Pyrolysis of carbonaceous fuels　　Fig. 23　Combustion and gasification of char

(5) In addition, the reversible gas phase water-gas shift reaction reaches equilibrium very fast at the temperatures in a gasifier. This balances the concentrations of carbon monoxide, steam, carbon dioxide and hydrogen, via the reaction $CO + H_2O \longleftrightarrow H_2 + CO_2$.

In essence, a limited amount of oxygen or air is introduced into the reactor to allow some of the organic material to be "burned" to produce carbon dioxide and energy, which drives a second reaction that further converts organic material to hydrogen and additional carbon dioxide. Further reactions occur when the formed carbon monoxide and residual water from the organic material react to form methane and excess carbon dioxide ($4CO + 2H_2O \longrightarrow CH_4 + 3CO_2$). This third reaction occurs more abundantly in reactors that increase the residence time of the reactive gases and organic materials, as well as temperature and pressure. Catalysts are used in more sophisticated reactors to improve reaction rates, thus moving the system closer to the reaction equilibrium for a fixed residence time.

6.3.3　Processes

Several types of gasifiers (shown in Fig. 24) are currently available for commercial use: counter-current fixed bed (up draft), co-current fixed bed (down draft), fluidized bed, entrained flow, plasma, and free radical.

(1) Counter-current fixed bed (Up draft) gasifier

A fixed bed of carbonaceous fuel (e.g. coal or biomass) through which the "gasification agent" (steam, oxygen and/or air) flows in counter-current configuration. The ash is either removed in the dry condi-

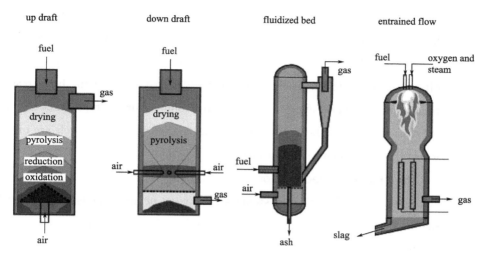

Fig. 24 Main gasifier types

tion or as a slag. The slagging gasifiers have a lower ratio of steam to carbon, achieving temperatures higher than the ash fusion point. The nature of the gasifier means that the fuel must have high mechanical strength and must ideally be non-caking in order to form a permeable bed, although recent developments have reduced these restrictions to some extent. The throughput for this type of gasifier is relatively low. It has high thermal efficiency due to the ralatively low temperatures in the gas outlet. However, this means that tar and methane production is significant at typical operation temperatures, so product gas must be extensively cleaned before use. The tar can be recycled into the reactor.

In the gasification of fine, undensified biomass such as rice hulls, it is necessary to blow air into the reactor by means of a fan. This creates very high gasification temperature, as high as 1000℃. Above the gasification zone, a bed of fine and hot char is formed, and as the gas is forced to blow through this bed, most complex hydrocarbons are broken down into simple components of hydrogen and carbon monoxide.

(2) Co-current fixed bed (Down draft) gasifier

Similar to the counter-current type, but the gasification agent gas flows in co-current configuration with the fuel (downwards, hence the name "down draft gasifier"). Heat needs to be added to the upper part of the bed, either by combusting small amounts of the fuel or from external heat sources. The produced gas leaves the gasifier at a high temperature, and most of this heat is often transferred to the gasification agent added in the top of the bed, resulting in an energy efficiency on level with the counter-current type. Since all tars must pass through a hot bed of char in this configuration, tar levels are much lower than the counter-current type.

(3) Fluidized bed reactor

The fuel is fluidized in steam and oxygen or air. The ash is removed dry or as heavy agglomerates that defluidize. The temperatures are relatively low in dry ash gasifiers, so the fuel must be highly reactive; and thus low-grade coals are particularly suitable. The agglomerating gasifiers have slightly higher temperatures, and higher rank coals are suitable. The fuel throughput is higher than the fixed bed, but not as high as the entrained flow gasifier. The conversion efficiency can be rather low due to elutriation of carbonaceous material. Recycle or subsequent combustion of solids can be used to increase conversion efficiency. Fluidized bed gasifiers are most useful for fuels that form highly corrosive ash that would damage the walls of slagging gasifiers. Biomass fuels generally contain high levels of corrosive ash.

(4) Entrained flow gasifier

A dry pulverized solid, an atomized liquid fuel or a fuel slurry is gasified with oxygen (much less frequent: air) in co-current flow. The gasification reactions take place in a dense cloud of very fine particles. Most coals are suitable for this type of gasifier because of the high operating temperatures and because the coal particles are well separated from one another.

The high temperatures and pressures also mean that a higher throughput can be achieved, however the thermal efficiency is somewhat lower as the gas must be cooled before it can be cleaned with existing technology. The high temperatures also mean that tar and methane are not present in the product gas; however the oxygen requirement is higher than for the other types of gasifiers. All entrained flow gasifiers remove the major part of the ash as a slag since the operating temperature is well above the ash fusion point.

A smaller fraction of the ash is produced either as a very fine dry fly ash or as a black-colored fly ash slurry. Some fuels, in particular certain types of biomasses, can form slag that is corrosive for ceramic inner walls that serve to protect the gasifier outer wall. However some entrained flow type of gasifiers do not possess a ceramic inner wall but have an inner water (or steam)-cooled wall covered with partially solidified slag. These types of gasifiers do not suffer from corrosive slags.

Some fuels have ashes with very high ash fusion points. In this case mostly limestone is mixed with the fuel prior to gasification. Addition of a little limestone will usually suffice for the lowering of the fusion temperatures. The fuel particles must be much smaller than for other types of gasifiers. This means the fuel must be pulverized, which requires somewhat more energy than for the other types of gasifiers. By far the most energy consumption related to entrained flow gasification is not the milling of the fuel but the production of oxygen used for the gasification.

(5) Plasma gasifier

In a plasma gasifier a high-voltage current is fed to a torch, creating a high-temperature arc. The inorganic residue is retrieved as a glass-like substance.

6.3.4　Feedstock

There is a large number of different feedstock types for use in a gasifier, each with different characteristics, including size, shape, bulk density, moisture content, energy content, chemical composition, ash fusion characteristics, and homogeneity of all these properties. Coal and petroleum coke are used as primary feedstocks for many large gasification plants worldwide. Additionally, a variety of biomass and waste-derived feedstocks can be gasified, with wood pellets and chips, waste wood, plastics and aluminium, municipal solid waste, refuse-derived fuel, agricultural and industrial wastes, sewage sludge, switchgrass, discarded seed corn, corn stover and other crop residues all being used.

ChemRec has developed a process for gasification of black liquor.

Waste disposal

Waste gasification (as shown in Fig. 25) has several advantages over incineration: ① The necessary extensive flue gas cleaning may be performed on the syngas instead of the much larger volume of flue gas after combustion. ② Electric power may be generated in engines and gas

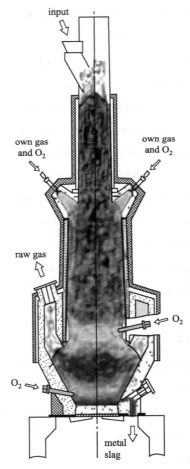

Fig. 25 HTCW reactor, one of several proposed waste gasification processes

Note: According to the sales and sales management consultants KBI Group, a pilot plant in Arnstadt implementing this process has completed initial tests

turbines, which are much cheaper and more efficient than the steam cycle used in incineration. Even though fuel cells may potentially be used, they have rather severe requirements regarding the purity of the gas. ③Chemical processing (Gas to liquids) of the syngas may produce other synthetic fuels instead of electricity. ④Some gasification processes treat ash containing heavy metals at very high temperatures so that it is released in a glassy and chemically stable form.

A major challenge for waste gasification technologies is to reach an acceptable (positive) gross electric efficiency. The high efficiency of converting syngas to electric power is counteracted by significant power consumption in the waste preprocessing, the consumption of large amounts of pure oxygen (which is often used as gasification agent), and gas cleaning. Another challenge becoming apparent when implementing the processes in real life is to obtain long service intervals in the plants, so that it is not necessary to close down the plant every few months for cleaning the reactor.

Environmental advocates have called gasification "incineration in disguise" and argue that the technology is still dangerous to air quality and public health. "Since 2003 numerous proposals for waste treatment facilities hoping to use... gasification technologies failed to receive final approval to operate when the claims of project proponents did not withstand public and governmental scrutiny of key claims," according to the Global Alliance for Incinerator Alternatives. One facility which operated from 2009 to 2011 in Ottawa had 29 "emission incidents" and 13 "spills" over those three years. It was also only able to operate roughly 25% of the time.

Several waste gasification processes have been proposed, but few have been built and tested, and only a handful of the processes have been implemented as plants processing real waste, and most of the time in combination with fossil fuels.

One plant (in Chiba, Japan, using the Thermoselect Process) has been processing industrial wastes since 2000, but has not yet documented positive net energy production from the process.

In the United States, gasification of waste is expanding across the country. Zegen is operating a waste gasification demonstration facility in New Bedford, Massachusetts. The facility was designed to demonstrate gasification of specific non-MSW waste streams using liquid metal gasification. This facility came after widespread public opposition shelved plans for a similar plant in Attleboro, Massachusetts. In addition, construction of a biomass gasification plant was approved in DeKalb County, Georgia on June 14, 2011.

Also in the USA, plasma is being used to gasify municipal solid waste, hazardous waste and biomedical waste at the Hurlburt Field Florida Special Operations Command Air Force base. PyroGenesis Canada Inc. is the technology provider. InnerPoint Energy in Cedar Hill, Missouri, has a running gasification system.

6.3.5 Current applications

Syngas can be used for heat production and for generation of mechanical and electric power. Like other gaseous fuels, producer gas gives greater control over power levels when compared to solid fuels, leading to more efficient and cleaner operation. Syngas can also be used for further processing into liquid fuels or chemicals.

(1) Heat

Gasifiers offer a flexible option for thermal applications, as they can be retrofitted into existing gas-fueled devices such as ovens, furnaces, boilers, etc., where syngas may replace fossil fuels. Heating values of syngas are generally around $4 \sim 10$ MJ/m^3.

(2) Electricity

Currently Industrial-scale gasification is primarily used to produce electricity from fossil fuels such as coal, where the syngas is burned in a gas turbine. Gasification is also used industrially in the production of electricity, ammonia and liquid fuels (oil) using Integrated Gasification Combined Cycles (IGCC), with the possibility of producing methane and hydrogen for fuel cells. IGCC is also a more efficient method of CO_2 capture as compared to conventional technologies. IGCC demonstration plants have been operating since the early 1970s and some of the plants constructed in the 1990s are now entering commercial service.

(3) Combined heat and power

In small business and building applications, where the wood source is sustainable, 250~1000 kW electricity and new zero carbon biomass gasification plants have been installed in Europe that produce tar free syngas from wood and burn it in reciprocating engines connected to a generator with heat recovery. This type of plant is often referred to as a wood biomass combined heat and power station (CHP) unit but is a plant with seven different processes: biomass processing, fuel delivery, gasification, gas cleaning, waste disposal, electricity generation and heat recovery.

(4) Transport fuel

Diesel engines can be operated on dual fuel mode using producer gas. Diesel substitution of over 80% at high loads and 70%~80% under normal load variations can easily be achieved. Spark ignition engines and solid oxide fuel cells can operate on 100% gasification gas. Mechanical energy from the engines may be used for driving water pumps for irrigation or for coupling with an alternator for electric power generation.

While small scale gasifiers have existed for well over 100 years, there are few sources to obtain a ready-to-use machine. Small scale devices are typically DIY projects. However, currently in the United States, several companies offer gasifiers to operate small engines.

(5) Renewable energy and fuels

In principle, gasification can proceed from just about any organic material, including biomass and plastic waste. The resulting syngas can be combusted. Alternatively, if the syngas is clean enough, it may be used for power production in gas engines, gas turbines or even fuel cells, or converted efficiently to dimethyl ether by methanol dehydration, methane via the Sabatier reaction, or diesel-like synthetic fuel via the Fischer-Tropsch synthesis. In many gasification processes most of the inorganic components of the input material, such as metals and minerals, are retained in the ash. In some gasification processes (slagging gasification) the ash has the form of a glassy solid with low leaching properties, but the net power production in slagging gasification is low (sometimes negative) and the costs are higher.

Regardless of the final fuel form, gasification itself and subsequent processing neither directly emits nor traps greenhouse gases such as carbon dioxide. Power consumption in the gasification and syngas conversion processes may be significant though, and may indirectly cause CO_2 emissions; in slagging and plasma gasification, the electricity consumption may even exceed any power production from the syngas.

Combustion of syngas or derived fuels emits exactly the same amount of carbon dioxide as would have been emitted from direct combustion of the initial fuel. Biomass gasification and combustion could play a significant role in a renewable energy economy, since biomass production removes the same amount of CO_2 from the atmosphere as is emitted from gasification and combustion. While other biofuel technologies such as biogas and biodiesel are carbon neutral, gasification in principle may run on a wider variety of input materials and can be used to produce a wider variety of output fuels.

There are at present a few industrial scale biomass gasification plants. Since 2008 in Svenljunga, Sweden, a biomass gasification plant generates energy up to 14 MW of heating energy, supplying industries and citizens of Svenljunga with process steam and district heating, respectively. The gasifier uses biomass fuels such as CCA (or creosote)-impregnated waste wood and other kinds of recycled wood to produce syngas that is combusted on site. In 2011 a similar gasifier, using the same kinds of fuels, was installed at Munkfors Energy's CHP plant. The CHP plant will generate 2 MW of electrical energy and 8 MW of district heating energy.

Chapter 7 Anaerobic digestion for municipal solid waste

Anaerobic digestion is a collection of processes by which microorganisms break down biodegradable material in the absence of oxygen. The process is used for industrial or domestic purposes to manage waste or to produce fuels. Much of the fermentation used industrially to produce food and drink products, as well as home fermentation, uses anaerobic digestion.

Anaerobic digestion occurs naturally in some soils, lake and oceanic basin sediments, where it is usually referred to as "anaerobic activity". This is the source of marsh gas methane as discovered by Alessandro Volta in 1776.

The digestion process begins with bacterial hydrolysis of the input materials. Insoluble organic polymers, such as carbohydrates, are broken down to soluble derivatives that become available for other bacteria. Acidogenic bacteria then convert the sugars and amino acids into carbon dioxide, hydrogen, ammonia, and organic acids. These bacteria convert the resulting organic acids into acetic acid, along with additional ammonia, hydrogen, and carbon dioxide. Finally, methanogens convert these products to methane and carbon dioxide. The methanogenic archaea populations play an indispensable role in anaerobic wastewater treatments.

Anaerobic digestion is used as part of the process to treat biodegradable waste and sewage sludge. As part of an integrated waste management system, anaerobic digestion reduces the emission of landfill gas into the atmosphere. Anaerobic digesters can also be fed with specially grown energy crops, such as maize.

Anaerobic digestion is widely used as a source of renewable energy. The process produces a biogas, consisting of methane, carbon dioxide and trace amount of other "contaminant" gases. This biogas can be used directly as fuel in combined heat and power gas engines, or upgraded to biomethane of natural gas quality. The additional nutrient-rich digestate produced can be used as fertilizer.

With the re-use of waste as a resource and new technological approaches that have lowered capital costs, anaerobic digestion has in recent years received increased attention among governments in a number of countries, including the United Kingdom.

7.1 Description

Anaerobic digestion is the biological decomposition of organic material in the absence of oxygen to produce biogas. The major products of this process are methane and carbon dioxide. The other byproducts include water vapor, ammonia and hydrogen sulfide. A literature search shows that manure (of cow or chicken) could produce 60% methane, 35% carbon dioxide and 5% others in composition through anaerobic digestion. The biological decomposition of the organic material (i.e. waste) involves a series of chemical reactions. The reaction below shows a general breakdown of waste into the final major product methane (Fig. 26).

The waste contains carbohydrates, proteins, fats, and others. First, acidogens converts waste to volatile fatty acids (VFA), and then VFA can be converted to methane as a major product by methanogens (Fig. 27). The production of methane could be affected by factors such as temperature, the type of waste, hydraulic retention time (the time waste spent in the digesters to produce methane), pH, the percent of solids, and pretreatment.

How it works...

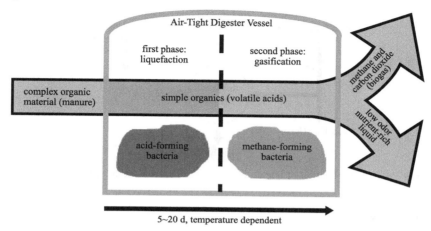

Fig. 26　A general breakdown of waste into the final major product methane

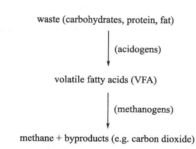

Fig. 27　Products during breakdown of waste into the final major product methane

Waste like cow manure can be fed to anerobic digester, and the biogas generated from the digester can be used to fuel an engine, which will drive a generator to produce electricity. The waste heat can be recovered, and used to heat the water and fed to the digester to enhance the reaction. The hot water fed into the digester should not have the temperature that could stop the digestion process. Methanogens are found to reproduce in extreme environments from 15℃ to 100℃. Hence, the hot water that is fed into the digester should not exceed this range in order to keep the digester running and producing biogas for sustainable energy. An application of anaerobic digestion is shown in Fig. 28.

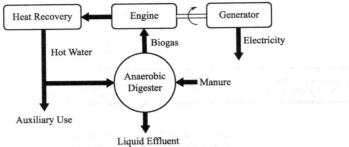

Fig. 28 An application of anaerobic digestion for sustainable energy production

7.2 Process

Many microorganisms affect anaerobic digestion, including acetic acid-forming bacteria (acetogens) and methane-forming archaea (methanogens). These organisms promote a number of chemical processes in converting the biomass to biogas.

Gaseous oxygen is excluded from the reactions by physical containment. Anaerobes utilize electron acceptors from sources other than oxygen gas. These acceptors can be the organic material itself or may be supplied by inorganic oxides from within the input material. When the oxygen source in an anaerobic system is derived from the organic material itself, the "intermediate" end products are primarily alcohols, aldehydes, and organic acids, plus carbon dioxide. In the presence of specialised methanogens, the intermediates are converted to the "final" end products of methane, carbon dioxide, and trace levels of hydrogen sulfide. In an anaerobic system, the majority of the chemical energy contained within the starting material is released by methanogenic bacteria as methane.

Populations of anaerobic microorganisms typically take a significant period of time to establish themselves to be fully effective. Therefore, common practice is to introduce anaerobic microorganisms from materials with existing populations, a process known as "seeding" the digesters, typically accomplished with the addition of sewage sludge or cattle slurry.

7.2.1 Process stages

The four key stages of anaerobic digestion involve hydrolysis, acidogenesis, acetogenesis and methanogenesis. The overall process can be described by the chemical reaction below, where organic material such as glucose is biochemically digested into carbon dioxide (CO_2) and methane (CH_4) by the anaerobic microorganisms.

$$C_6H_{12}O_6 \longrightarrow 3CO_2 + 3CH_4$$

(1) Hydrolysis

In most cases, biomass is made up of large organic polymers. In order for the bacteria in anaerobic digesters to access the potential energy of the material, these chains must first be broken down into smaller constituent parts. These constituent parts, or monomers, such as sugars, are readily available to other bacteria. The process of breaking these chains and dissolving the smaller molecules into solution is called hydrolysis. Therefore, hydrolysis of these high-molecular-weight polymeric components is the necessary first step in anaerobic digestion. Through hydrolysis the complex organic molecules are broken down into simple sugars, amino acids, and fatty acids.

Acetate and hydrogen produced in the first stages can be used directly by methanogens. Other molecules, such as VFAs with a chain length greater than that of acetate must first be catabolized into compounds that can be directly used by methanogens.

(2) Acidogenesis

The biological process of acidogenesis results in further breakdown of the remaining components by acidogenic (fermentative) bacteria. Here, VFAs are created, along with ammonia, carbon dioxide, and hydrogen sulfide, as well as other byproducts. The process of acidogenesis is similar to the way milk sours.

(3) Acetogenesis

The third stage of anaerobic digestion is acetogenesis. Here, simple molecules created through the acidogenesis phase are further digested by acetogens to produce largely acetic acid, as well as carbon dioxide and hydrogen.

(4) Methanogenesis

The terminal stage of anaerobic digestion is the biological process of methanogenesis. Here, methanogens use the intermediate products of the preceding stages and convert them into methane, carbon dioxide, and water. These components make up the majority of the biogas emitted from the system. Methanogenesis is sensitive to both high and low pH and occurs between pH 6.5 and pH 8. The remaining, indigestible material the microbes cannot use and any dead bacterial residues constitute the digestate.

7.2.2 Configuration

Anaerobic digesters can be designed and engineered to operate using a number of different configurations and can be categorized into batch vs. continuous process mode, mesophilic vs. thermophilic temperature conditions, high vs. low portion of solids, and single stage vs. multistage processes. More initial construction funds and a larger volume of the batch digester is needed to handle the same amount of waste as a continuous process digester. More heat energy is demanded in a thermophilic system compared to a mesophilic system and the former has a larger gas output capacity and higher methane gas content. Low solids content will handle up to 15% solid content. Configuration with solid portion above this level is considered high solids content and can also be known as dry digestion. In a single stage process, one reactor contains the four steps of anaerobic digestion while a multistage process utilizes two or more reactors for digestion to separate the methanogenesis and hydrolysis phases.

(1) Batch or continuous

Anaerobic digestion can be performed as a batch process or a continuous process. In a batch system, biomass is added to the reactor at the start of the process. The reactor is then sealed for the duration of the process. In its simplest form, batch processing needs inoculation with already processed material to start the anaerobic digestion. In a typical scenario, biogas production will be formed with a normal distribution pattern over time. Operators can use this fact to determine when they believe the digestion process of the organic material has completed. There can be severe odour issues if a batch reactor is opened and emptied before the process is well completed. A more advanced type of batch approach has limited the odour issues by integrating anaerobic digestion with in-vessel composting. In this approach inoculation takes place through the use of recirculated degasified percolate. After anaerobic digestion is completed, the biomass is kept in the reactor which is then used for in-vessel composting before the reactor is opened. As the batch digestion is simple and requires less equipment and lower levels of design effort, it is typically a cheaper form of digestion. The use of more than one batch reactor at a plant can ensure constant production of biogas.

In continuous digestion processes, organic matter is constantly added (continuous and completely mixed) or added in stages to the reactor (continuous plug flow; first in-first out). Here, the end products are constantly or periodically removed, resulting in constant production of biogas. A single or multiple digesters in sequence may be used. Examples of this form of anaerobic digestion include continuous stirred tank reactors, upflow anaerobic sludge blankets, expanded granular sludge beds and internal circulation reactors.

(2) Temperature

The two conventional operational temperature levels for anaerobic digesters determine the species of methanogens in the digesters: (1) Mesophilic digestion takes place optimally around 30℃ to 38℃, or at ambient temperatures between 20℃ and 45℃, where mesophiles are the primary microorganisms present. (2) Thermophilic digestion takes place optimally around 49℃ to 57℃, or at elevated temperatures up to 70℃, where thermophiles are the primary microorganisms present.

A limit case has been reached in Bolivia, with anaerobic digestion in working conditions of below 10℃. The anaerobic process is very slow, taking more than three times the normal mesophilic process. In experimental work at the University of Alaska Fairbanks, a 1,000 L digester using psychrophiles harvested from "mud from a frozen lake in Alaska" has produced 200~300 L of methane per day, about 20% to 30% of the output from digesters in warmer climates. Mesophilic species outnumber thermophiles, and they are also more tolerant to changes in environmental conditions than thermophiles. Mesophilic systems are, therefore, considered to be more stable than thermophilic digestion systems. In contrast, while thermophilic digestion systems are considered less stable, their energy input is higher, with more biogas being removed from the organic matter in an equal amount of time. The increased temperatures facilitate faster reaction rates, and thus faster gas yields. Operation at higher temperatures facilitates greater pathogen reduction of the digestate. In countries where legislation, such as the Animal By-Products Regulations in the European Union, requires digestate to meet certain levels of pathogen reduction there may be a benefit using thermophilic temperatures instead of mesophilic.

Additional pretreatment can be used to reduce the necessary retention time to produce biogas. For example, certain processes can be used to shred the substrates to increase the surface area or a thermal pretreatment stage (such as pasteurisation) to significantly enhance the biogas output. The pasteurisation process can also be used to reduce the pathogenic concentration in the digestate leaving the anaerobic digester. Pasteurisation may be achieved by heat treatment combined with maceration of the solids.

(3) Solids content

In a typical scenario, three different operational parameters are associated with the solids content of the feedstock to the digesters: (1) High solids (dry-stackable substrate); (2) High solids (wet-pumpable substrate); (3) Low solids (wet-pumpable substrate).

High solids (dry-stackable substrate) digesters are designed to process materials with a solids content between 25% and 40%. Unlike wet digesters that process pumpable slurries, high solids (dry-stackable substrate) digesters are designed to process solid substrates without the addition of water. The primary styles of dry digesters are continuous vertical plug flow and batch tunnel horizontal digesters. Continuous vertical plug flow digesters are upright, cylindrical tanks where feedstock is continuously fed into the top of the digester, and flows downward by gravity during digestion. In batch tunnel digesters, the feedstock is deposited in tunnel-like chambers with a gas-tight door. Neither approach has mixing inside the digester. The amount of pretreatment, such as contaminant removal, depends both upon the nature of the waste streams being processed and the desired quality of the digestate. Size reduction (grinding) is beneficial in continuous vertical systems, as it accelerates digestion, while batch systems avoid grinding and instead require structure (e.g. yard waste) to reduce compaction of the stacked pile. Continuous vertical dry digesters have a smaller footprint due to the shorter effective retention time and vertical design. Wet digesters can be designed to operate in either a high solids content, with a total suspended solids (TSS) concentration greater than $\sim 20\%$, or a low solids concentration less than $\sim 15\%$.

High solids (wet) digesters process a thick slurry that requires more energy input to move and process the feedstock. The thickness of the material may also lead to associated problems with abrasion. High solids digesters will typically have a lower land requirement due to the lower volumes associated with the moisture. High solids digesters also require correction of conventional performance calculations (e.g. gas production, retention time, kinetics, etc.) originally based on very dilute sewage digestion concepts, since larger fractions of the feedstock mass are potentially convertible to biogas.

Low solids (wet) digesters can transport material through the system using standard pumps that require significantly lower energy input. Low solids digesters require a larger amount of land than high solids due to the increased volumes associated with the increased liquid-to-feedstock ratio of the digesters. There are benefits associated with operation in a liquid environment, as it enables more thorough circulation of materials and contact between the bacteria and their food. This enables the bacteria to more readily access the substances on which they are feeding, and increases the rate of gas production.

(4) Complexity

Digestion systems can be configured with different levels of complexity. In a single-stage digestion system (one-stage), all of the biological reactions occur within a single, sealed reactor or holding tank. Using a single stage reduces construction costs, but results in less control of the reactions occurring within the system. Acidogenic bacteria reduce the pH of the tank through the production of acids. Methanogenic bacteria, as outlined earlier, operate in a strictly defined pH range. Therefore, the biological reactions of the different species in a single-stage reactor can be in direct competition with each other. Another one-stage reaction system is an anaerobic lagoon. These lagoons are pond-like, earthen basins used for the treatment and long-term storage of manures. Here the anaerobic reactions are contained within the natural anaerobic sludge in the pool.

In a two-stage digestion system (multistage), different digestion vessels are optimised to bring maximum control over the bacterial communities living within the digesters. Acidogenic bacteria produce organic acids and grow and reproduce more quickly than methanogenic bacteria. Methanogenic bacteria require stable pH and temperature to optimise their performance.

Under typical circumstances, hydrolysis, acetogenesis, and acidogenesis occur within the first reaction vessel. The organic material is then heated to the required operational temperature (either mesophilic or thermophilic) prior to being pumped into a methanogenic reactor. The initial hydrolysis or acidogenesis tanks prior to the methanogenic reactor can provide a buffer to the rate at which feedstock is added. Some European countries require a degree of elevated heat treatment to kill harmful bacteria in the input waste. In this instance, there may be a pasteurisation or sterilisation stage prior to digestion or between the two digestion tanks. Notably, it is not possible to completely isolate the different reaction phases, and often some biogas is produced in the hydrolysis or acidogenesis tanks.

(5) Residence time

The residence time in a digester varies with the amount and type of feed material, and with the configuration of the digestion system. In a typical two-stage mesophilic digestion, residence time varies between 15 and 40d, while for a single-stage thermophilic digestion, residence time is normally shorter and takes around 14d. The plug-flow nature of some of these systems will mean the full degradation of the material may not have been realised in this timescale. In this event, digestate exiting the system will be darker in colour and will typically have more odour.

In the case of an upflow anaerobic sludge blanket digestion (UASB), hydraulic residence times can be as short as 1h to 1d, and solid retention times can be up to 90d. In this manner, a UASB system is able to separate solids and hydraulic retention times with the use of a sludge blanket. Continuous digesters have mechanical or hydraulic devices, depending on the level of solids in the material, to mix the contents, enabling the bacteria and the food to be in contact. They also allow excess material to be continuously extracted to maintain a reasonably constant volume within the digestion tanks.

(6) Inhibition

The anaerobic digestion process can be inhibited by several compounds, affecting one or more of the bacterial groups responsible for the degradation steps of different organic materials. The degree of the inhibition depends, among other factors, on the concentration of the inhibitor in the digester. Potential inhibitors are ammonia, sulfide, light metal ions (Na, K, Mg, Ca, Al), heavy metals, some organics (chlorophenols, halogenated aliphatics, N-substituted aromatics, long-chain fatty acids), etc.

7.2.3 Feedstocks

The most important initial issue when considering the application of anaerobic digestion systems is the feedstock to the process. Almost all organic materials can be processed with anaerobic digestion; however, if biogas production is the aim, the level of putrescibility is the key factor in its successful application. The more putrescible (digestible) the material, the higher the gas yields possible from the system.

Feedstocks can include biodegradable waste materials, such as waste paper, grass clippings, leftover food, sewage, and animal waste. Woody wastes are the exception, because they are largely unaffected by digestion, and most anaerobes are unable to degrade lignin. Xylophalgeous anaerobes (lignin consumers) or high-temperature pretreatment, such as pyrolysis, can be used to break down the lignin. Anaerobic digesters can also be fed with specially grown energy crops, such as silage, for dedicated biogas production. In Germany and continental Europe, these facilities are referred to as "biogas" plants. A co-digestion or co-fermentation plant is typically an agricultural anaerobic digester that accepts two or more types of input materials for simultaneous digestion.

The length of time required for anaerobic digestion depends on the chemical complexity of the material. Material rich in easily digestible sugars breaks down quickly whereas intact lignocellulosic material rich in cellulose and hemicellulose polymers can take much longer to break down. Anaerobic microorganisms are generally unable to break down lignin, the recalcitrant aromatic component of biomass.

Anaerobic digesters were originally designed for operation using sewage sludge and manures. Sewage and manure are not, however, the material with the most potential for anaerobic digestion, as the biodegradable part has already had much of the energy content taken out by the animals that produced it. Therefore, many digesters operate with co-digestion of two or more types of feedstock. For example, in a farm-based digester that uses dairy manure as the primary feedstock, the gas production may be significantly increased by adding a second feedstock, e. g., grass and corn (typical on-site feedstock), or various organic by-products, such as slaughterhouse waste, fats oils and grease from restaurants, organic household waste, etc. (typical off-site feedstock).

Digesters processing dedicated energy crops can achieve high levels of degradation and biogas production. Slurry-only systems are generally cheaper, but generate far less energy than those using crops, such as maize and grass silage; by using a modest amount of crop material (30%), an anaerobic digestion plant can increase the energy output tenfold for only three times the capital cost, relative to a slurry-only system.

(1) Substrate composition

Substrate composition is a major factor in determining the methane yield and methane production rates from the digestion of biomass. Techniques to determine the compositional characteristics of the feedstock are available, while parameters such as solids, elemental, and organic analyses are important for digester design and operation. Methane yield can be estimated from the elemental composition of substrate along with an estimate of its degradability (the fraction of the substrate that is converted to biogas in a reactor). In order to predict biogas composition (the relative fractions of methane and carbon dioxide), it is necessary to estimate carbon dioxide partitioning between the aqueous and gas phases, which requires additional information (reactor temperature, pH, and substrate composition) and a chemical speciation model.

(2) Moisture content

A second consideration related to the feedstock is moisture content. Drier, stackable substrates, such as food and yard waste, are suitable for digestion in tunnel-like chambers. Tunnel-style systems typically have near-zero wastewater discharge as well, so this style of system has advantages where the discharge of digester liquids is a liability. The wetter the material, the more suitable it will be to handling with standard pumps instead of energy-intensive concrete pumps and physical means of movement. Also, the wetter the material, the more volume and area it takes up relative to the levels of gas produced. The moisture content of the target feedstock will also affect what type of system is applied to its treatment. To use a high-solids anaerobic digester for dilute feedstocks, bulking agents such as compost, should be applied to increase the solids content of the input material. Another key consideration is the C/N ratio of the input material. This ratio is the balance of food a microbe requires to grow; the optimal C/N ratio is $(20 \sim 30):1$. Excess N can lead to ammonia inhibition of digestion.

(3) Contamination

The level of contamination of the feedstock material is a key consideration. If the feedstock to the digesters has significant levels of physical contaminants, such as plastic, glass, or metals, then processing to remove the contaminants will be required for the material to be used. If it is not removed, then the digesters can be blocked and will not function efficiently. It is with this understanding that mechanical biological treatment plants are designed. The higher the level of pretreatment a feedstock requires, the more processing machinery will be required, and, hence, the project will have higher capital costs.

After sorting or screening to remove any physical contaminants from the feedstock, the material is often shredded, minced, and mechanically or hydraulically pulped to increase the surface area available to microbes in the digesters and, hence, increase the speed of digestion. The maceration of solids can be achieved by using a chopper pump to transfer the feedstock material into the airtight digester, where anaerobic treatment takes place.

7.2.4 Applications

Using anaerobic digestion technologies can help to reduce the emission of greenhouse gases in a number of key ways: (1) Replacement of fossil fuels; (2) Reducing or eliminating the energy footprint of waste treatment plants; (3) Reducing methane emission from landfills; (4) Displacing industrially produced chemical fertilizers; (5) Reducing vehicle movements; (6) Reducing electrical grid transportation losses; (7) Reducing usage of LPG for cooking. A schematic of an anaerobic digester as part of a sanitation system is shown in Fig. 29. It produces a digested slurry (digestate) that can be used as a fertilizer, and biogas that can be used for energy.

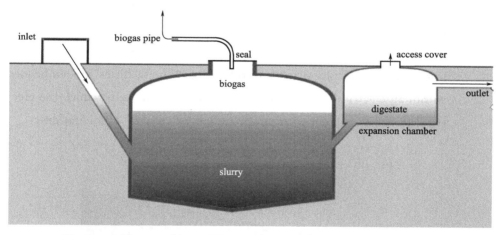

Fig. 29　Schematic of an anaerobic digester as part of a sanitation system

(1) Waste and wastewater treatment

Anaerobic digestion is particularly suited to organic material, and is commonly used for industrial effluent, wastewater and sewage sludge treatment (Fig. 30). Anaerobic digestion, a simple process, can greatly reduce the amount of organic matter which might otherwise be destined to be dumped at sea, dumped in landfills, or burnt in incinerators.

Pressure from environmentally related legislation on solid waste disposal methods in developed countries has increased the application of anaerobic digestion as a process for reducing waste volumes and generating useful byproducts. It may either be used to process the source-separated fraction of municipal waste or alternatively combined with mechanical sorting systems, to process residual mixed municipal waste. These facilities are called mechanical biological treatment plants.

If the putrescible waste processed in anaerobic digesters were disposed of in a landfill, it would break down naturally and often anaerobically. In this case, the gas will eventually escape into the atmosphere. As methane is about 20 times more potent as a greenhouse gas than carbon dioxide, this has significantly negative impacts on environment.

In countries that collect household waste, the use of local anaerobic digestion facilities can help to reduce the amount of waste that requires transportation to centralized landfill sites or incineration facilities. This reduced burden on transportation reduces carbon emissions from the collection vehicles. If localized anaerobic digestion facilities are embedded within an electrical distribution network, they can help reduce the electrical losses associated with power transmission over a national grid.

Fig. 30 Anaerobic digesters in a sewage treatment plant where the methane gas is managed by burning through a gas flare

(2) Power generation

In developing countries, simple home-and farm-based anaerobic digestion systems offer the potential for low-cost energy for cooking and lighting. From 1975, China and India have both had large, government-backed schemes for adaptation of small biogas plants for use in the household for cooking and lighting. At present, projects for anaerobic digestion in the developing world can gain financial support through the United Nations Clean Development Mechanism if they are able to show they have reduced carbon emissions.

Methane and power produced in anaerobic digestion facilities can be used to replace energy derived from fossil fuels, and hence reduce emissions of greenhouse gases, because the carbon in biodegradable material is part of a carbon cycle. The carbon released into the atmosphere from the combustion of biogas has been removed by plants to grow in the recent past, usually within the last decade, but more typically within the last growing season. If the plants are regrown, taking the carbon out of the atmosphere once more, the system will be carbon neutral. In contrast, carbon in fossil fuels has been sequestered in the earth for many millions of years, the combustion of which increases the overall level of carbon dioxide in the atmosphere.

Biogas from sewage sludge treatment is sometimes used to run a gas engine to produce electricity, some or all of which can be used to run the sewage works. Some waste heat from the engine is then used to heat the digester. The waste heat is, in general, enough to heat the digester to the required temperatures. The power potential from sewage works is limited—in the UK, there are about 80 MW total of such generation, with the potential to increase to 150 MW, which is insignificant compared to the average power demand in the UK of about 35,000 MW. The scope for biogas generation from non-sewage waste biological matter—energy crops, food waste, abattoir waste, etc.—is much higher, estimated to be capable of about 3,000 MW. Farm biogas plants using animal waste and energy crops are expected to contribute to reducing CO_2 emissions and strengthen the grid, while providing UK farmers with additional revenues.

Some countries offer incentives in the form of, for example, feed-in tariffs for feeding electricity onto the power grid to subsidize green energy production.

In Oakland, California at the main wastewater treatment plant in East Bay Municipal Utility District (EBMUD), food waste is currently co-digested with primary and secondary municipal wastewater solids and other high-strength wastes. Compared to municipal wastewater solids digestion alone, food waste co-digestion has many benefits. Anaerobic digestion of food waste pulp from the EBMUD food waste process provides a higher normalized energy benefit than municipal wastewater solids: 730~1,300 kWh per ton of dry food waste applied compared to 560~940 kWh per ton of dry municipal wastewater solids applied.

(3) Grid injection

Biogas grid-injection is the injection of biogas into the natural gas grid. The raw biogas has to be previously upgraded to biomethane. This upgrading implies the removal of contaminants such as hydrogen sulphide or siloxanes, as well as the carbon dioxide. Several technologies are available for this purpose, the most widely implemented being pressure swing adsorption, water or amine scrubbing (absorption processes) and, in recent years, membrane separation. As an alternative, the electricity and the heat can be used for on-site generation, resulting in a reduction of losses in the transportation of energy. Typical energy losses in natural gas transmission systems range from 1% to 2%, whereas the current energy losses on a large electrical system range from 5% to 8%.

In October 2010, Didcot Sewage Works became the first in the UK to produce biomethane gas supplied to the national grid, for use in up to 200 homes in Oxfordshire. By 2017, UK electricity firm Ecotricity planned to have digesters fed by locally sourced grass to fuel 6,000 homes.

(4) Vehicle fuel

After upgrading with the above-mentioned technologies, the biogas (transformed into biomethane) can be used as vehicle fuel in adapted vehicles. This use is very extensive in Sweden, where over 38,600 gas vehicles exist, and 60% of the vehicle gas is biomethane generated in anaerobic digestion plants.

(5) Fertiliser and soil conditioner

The solid, fibrous component of the digested material can be used as a soil conditioner to increase the organic content of soils. Digester liquids can be used as a fertiliser to supply vital nutrients to soils instead of chemical fertilisers that require large amounts of energy to produce and transport. The use of manufactured fertilisers is, therefore, more carbon-intensive than anaerobic digester liquor fertiliser. In countries such as Spain, where many soils are organically depleted, the markets for the digested solids can be equally as important as the biogas.

(6) Cooking gas

By using a bio-digester, which produces the bacteria required for decomposing, cooking gas is generated. Organic garbage like fallen leaves, kitchen waste, food waste etc. are fed into a crusher unit, where the mixture is conflated with a small amount of water. The mixture is then fed into the bio-digester, where the bacteria decomposes it to produce cooking gas. This gas is piped to kitchen stove. A 2 m^3 of bio-digester can produce 2 m^3 of cooking gas. This is equivalent to 1 kg of LPG. The notable advantage of using a bio-digester is the sludge which is a rich organic manure.

7.2.5 Products

(1) Biogas

Biogas is the ultimate waste product of the bacteria feeding off the input biodegradable feedstock (the methanogenesis stage of anaerobic digestion is performed by archaea, a microorganism on a distinctly different branch of the phylogenetic tree of life to bacteria), and is mostly methane and carbon dioxide, with a small amount of hydrogen and trace hydrogen sulfide. (As produced, biogas also contains water vapor, with the fractional water vapor volume a function of biogas temperature). Most of the biogas is produced during the digestion, after the bacterial population has grown, and tapers off as the putrescible material is exhausted. The gas is normally stored on top of the digester in an inflatable gas bubble or extracted and stored next to the facility in a gas holder. Typical composition of biogas is shown in Table 21.

Table 21 Typical composition of biogas

Compound	Formula	Content/%
Methane	CH_4	50~75
Carbon dioxide	CO_2	25~50
Nitrogen	N_2	0~10
Hydrogen	H_2	0~1
Hydrogen sulfide	H_2S	0~3
Oxygen	O_2	0

The methane in biogas can be burned to produce both heat and electricity, usually with a reciprocating engine or microturbine often in a co-generation arrangement where the electricity and waste heat generated are used to warm the digesters or to heat buildings. Excess electricity can be sold to suppliers or put into the local grid. Electricity produced by anaerobic digesters is considered to be renewable energy and may attract subsidies. Biogas does not contribute to increasing atmospheric carbon dioxide concentrations because the gas is not released directly into the atmosphere and the carbon dioxide comes from an organic source with a short carbon cycle.

Biogas may require treatment or "scrubbing" to refine for use as a fuel. Hydrogen sulfide, a toxic product formed from sulfates in the feedstock, is released as a trace component of the biogas. National environmental enforcement agencies, such as the U.S. Environmental Protection Agency or the English and Welsh Environment Agency, put strict limits on the levels of gases containing hydrogen sulfide in the gas, and, if the levels of hydrogen sulfide in the gas are high, gas scrubbing and cleaning equipment (such as amine gas treating) will be needed to process the biogas to within regionally accepted levels. Alternatively, the addition of ferrous chloride $FeCl_2$ to the digestion tanks inhibits hydrogen sulfide production.

Volatile siloxanes can also contaminate the biogas; such compounds are frequently found in household waste and wastewater. In digestion facilities accepting these materials as a component of the feedstock, low-molecular-weight siloxanes volatilise into biogas. When this gas is combusted in a gas engine, turbine, or boiler, siloxanes are converted into silicon dioxide (SiO_2), which deposits internally in the machine, increasing wear and tear. Practical and cost-effective technologies to remove siloxanes and other biogas contaminants are available at the present time. In certain applications, in-situ treatment can be used to increase the purity of methane by reducing the carbon dioxide content in the exhaust gas, purging the majority of it in a secondary reactor.

In countries such as Switzerland, Germany, and Sweden, the methane in the biogas may be compressed to be used as a vehicle fuel or input directly into the gas mains. In countries where the driver factor for the use of anaerobic digestion are renewable electricity subsidies, this route of treatment is less likely, as energy is required in this processing stage and the treatment reduces the overall levels available to sell.

(2) Digestate

Digestate is the solid remnants of the original input material to the digesters that the microbes cannot use. It also consists of the mineralised remains of the dead bacteria from within the digesters. Digestate can come in three forms: fibrous, liquid, or a sludge-based combination of the two fractions. In two-stage systems, different forms of digestate come from different digestion tanks. In single-stage digestion systems, the two fractions will be combined and, if desired, separated by further processing.

The second byproduct (acidogenic digestate) is a stable, organic material consisting largely of lignin and cellulose, but also a variety of mineral components in a matrix of dead bacterial cells; some plastic may be present. The material resembles domestic compost and can be used as such or to make low-grade building products, such as fibreboard. The solid digestate can also be used as feedstock for ethanol production.

The third byproduct is a liquid (methanogenic digestate) rich in nutrients, which can be used as a fertiliser, depending on the quality of the material being digested. Levels of potentially toxic elements (PTEs) should be chemically assessed. This will depend upon the quality of the original feedstock. In the case of most clean and source-separated biodegradable waste streams, the levels of PTEs will be low. In the case of wastes originating from industry, the levels of PTEs may be higher and will need to be taken into consideration when determining a suitable end use for the material.

Digestate typically contains elements, such as lignin, that cannot be broken down by the anaerobic microorganisms. Also, the digestate may contain ammonia that is phytotoxic, and may hamper the growth of plants if it is used as a soil-improving material. For these two reasons, a maturation or composting stage may be used after digestion. Lignin and other materials are available for degradation by aerobic microorganisms, such as fungi, helping reduce the overall volume of the material for transport. During this maturation, the ammonia will be oxidized into nitrates, improving the fertility of the material and making it more suitable as a soil improver. Large composting stages are typically used by dry anaerobic digestion technologies.

(3) Wastewater

The final output from anaerobic digestion systems is water, which originates from both the moisture content of the original waste that was treated and the water produced during the microbial reactions in the digestion systems. This water may be released by the dewatering of the digestate or may be implicitly separated from the digestate.

The wastewater exiting in the anaerobic digestion facility will typically have elevated levels of biochemical oxygen demand (BOD) and chemical oxygen demand (COD). These measures of the reactivity of the effluent indicate an ability to pollute. Some of this material is termed "hard COD", meaning it cannot be accessed by the anaerobic bacteria for conversion into biogas. If this effluent were put directly into watercourses, it would affect them negatively by causing eutrophication. As such, further treatment of the wastewater is often required. This treatment will typically be an oxidation stage where air is passed through the water in a sequencing batch reactors or reverse osmosis unit.

7.2.6 History

Reported scientific interest in the manufacturing of gas produced by the natural decomposition of organic matter dates from the 17th century, when Robert Boyle (1627—1691) and Stephen Hales (1677—1761) noted that disturbing the sediment of streams and lakes released flammable gas. In 1808 Sir Humphry Davy proved the presence of methane in the gases produced by cattle manure. In 1859 a leper colony in Bombay in India built the first anaerobic digester. In 1895, the technology was developed in Exeter, England, where a septic tank was used to generate gas for the sewer gas destructor lamp, a type of gas lighting. Also in England, in 1904, the first dual-purpose tank for both sedimentation and sludge treatment was installed in Hampton, London. In 1907, in Germany, a patent was issued for the Imhoff tank, an early form of digester.

Research on anaerobic digestion began in earnest in the 1930s.

Chapter 8 Composting for municipal solid waste

Compost is organic matter that has been decomposed and recycled as a fertilizer and soil amendment. Compost is a key ingredient in organic farming.

At the simplest level, the process of composting requires making a heap of wet organic matter known as green waste (leaves, food waste) and waiting for the materials to break down into humus after weeks or months. At present, methodical composting is a multi-step, closely monitored process with measured inputs of water, air, and carbon-and nitrogen-rich materials. The decomposition process is aided by shredding the plant material, adding water and ensuring proper aeration by regularly rotating the mixture. Worms and fungi further break down the material. Bacteria requiring oxygen to function (aerobic bacteria) and fungi manage the chemical process by converting the inputs into heat, carbon dioxide, and ammonium. The ammonium (NH_4^+) is the form of nitrogen used by plants. When available ammonium is not used by plants it is further converted by bacteria into nitrates (NO_3^-) through the process of nitrification. A community-level composting plant in a rural area in Germany is shown in Fig. 31.

Compost is rich in nutrients. It is used in gardens, landscaping, horticulture, and agriculture. The compost itself is beneficial for the land in many ways, including as a soil conditioner, a fertilizer, addition of vital humus or humic acids, and as a natural pesticide for soil. In ecosys-

Fig. 31 A community-level composting plant in a rural area in Germany

tems, compost is useful for erosion control, land and stream reclamation, wetland construction, and as landfill cover (see compost uses). Organic ingredients intended for composting can alternatively be used to generate biogas through anaerobic digestion.

8.1 Terminology

Composting of waste is an aerobic (in the presence of air) method of decomposing solid wastes. The process involves decomposition of organic waste into humus known as compost which is a good fertilizer for plants. However, the term "composting" is used worldwide with different meanings. Some textbooks on composting narrowly define composting as an aerobic form of decomposition, primarily by aerobic or facultative microbes. An alternative form of organic decomposition to composting is "anaerobic digestion".

For many people, composting is used to refer to several different types of biological processes. In North America, "anaerobic composting" is still a common term for what much of the rest of the world and in technical publications people call "anaerobic digestion". The microbes used and the processes involved are quite different between composting and anaerobic digestion.

8.2 Fundamentals

8.2.1 Carbon, nitrogen, oxygen, water

Composting organisms require four equally important ingredients to work effectively: (1) Carbon—for energy. the microbial oxidation of carbon produces the heat, if included at suggested levels. High carbon materials tend to be brown and dry. (2) Nitrogen—to grow and reproduce more organisms to oxidize the carbon. High nitrogen materials tend to be green (or colorful, such as fruits and vegetables) and wet. (3) Oxygen—for oxidizing the carbon in the decomposition process. (4) Water—with the right amounts to maintain activity without causing anaerobic conditions.

Certain ratios of these materials will provide beneficial bacteria with the nutrients to work at a rate that will heat up the pile. In that process much water will be released as vapor ("steam"), and the oxygen will be quickly depleted, explaining the need to manage the pile actively. The hotter the pile gets, the more often addition of air and water is necessary; the air/water balance is critical to maintaining high temperatures (135~160 °F/ 50~70℃) until the materials are broken down. At the same time, too much air or water also slows the process, as does too much carbon (or too little nitrogen). Hot container composting focuses on retaining the heat to increase decomposition rate and produce compost more quickly.

The most efficient composting occurs with an optimal C/N ratio of about 10 : 1 to 20 : 1. Rapid composting is favored by having a C/N ratio of ~30 or less. Theoretical analysis is confirmed by field tests that the substrate with a C/N ratio above 30 is nitrogen starved, and the substrate with a ratio below 15 is likely to emit a portion of nitrogen as ammonia.

Nearly all plant and animal materials have both carbon and nitrogen, but the amounts vary widely, with characteristics noted above (dry/wet, brown/green). Fresh grass clippings have an average ratio of about 15∶1 and dry autumn leaves about 50∶1 depending on species. Mixing equal parts by volume approximates the ideal C/N range. Few individual situations will provide the ideal mix of materials at any point. Observation of amounts, and consideration of different materials as a pile are built over time, by which a workable technique for the individual situation can be achieved quickly.

8.2.2 Microorganisms

With the proper mixture of water, oxygen, carbon, and nitrogen, microorganisms are able to break down organic matter to produce compost. The composting process is dependent on microorganisms to break down organic matter into compost. There are many types of microorganisms found in active compost among which the most common are: (1) Bacteria—The most numerous of all the microorganisms found in compost. Depending on the phase of composting, mesophilic or thermophilic bacteria may predominate. (2) Actinobacteria—Necessary for breaking down products such as newspaper, bark, etc. (3) Fungi—molds and yeast help break down materials that bacteria cannot, especially lignin in woody material. (4) Protozoa—Help consume bacteria, fungi and micro organic particulates. (5) Rotifers—Rotifers help control populations of bacteria and small protozoans.

In addition, earthworms not only ingest partly composted material, but also continually re-create aeration and drainage tunnels as they move through the compost.

The lack of a healthy microorganism community is the main reason why composting processes are slow in landfills with environmental factors such as lack of oxygen, nutrients or water being the cause of the depleted biological community.

8.2.3 Phases of composting

Under ideal conditions, composting proceeds through three major phases: (1) An initial, mesophilic phase, in which the decomposition is carried out under moderate temperatures by mesophilic microorganisms. (2) As the temperature rises, a second, thermophilic phase starts, in which the decomposition is carried out by various thermophilic bacteria under high temperatures. (3) As the supply of high-energy compounds dwindles, the temperature starts to fall, and the mesophiles once again predominate in the maturation phase.

8.2.4 Slow and rapid composting

There are many modern proponents of rapid composting that attempt to correct some of the perceived problems associated with traditional, slow composting. Many advocate that compost can be made in 2~3 weeks. Many such short processes involve a few changes to traditional methods, including smaller, more homogenized pieces in the compost, controlling carbon-to-nitrogen ratio (C/N) at 30 : 1 or less, and monitoring the moisture level more carefully. However, none of these parameters differ significantly from the early writings of compost researchers, suggesting that in fact modern composting has not made significant advances over the traditional methods that take a few months to work. For this reason and others, many modern scientists who deal with carbon transformations are sceptical that there is a "super-charged" way to get nature to make compost rapidly.

Both sides may be right to some extent. The bacterial activity in rapid high heat methods breaks down the material to the extent that pathogens and seeds are destroyed, and the original feedstock is unrecognizable. At this stage, the compost can be used to prepare fields or other planting areas. However, most professionals recommend that the compost be given time to cure before used in a nursery for starting seeds or growing young plants. The curing time allows fungi to continue the decomposition process and eliminate phytotoxic substances.

8.2.5 Pathogen removal

Composting can destroy pathogens or unwanted seeds. Unwanted living plants (or weeds) can be discouraged by covering with mulch/compost. The "microbial pesticides" in compost may include thermophiles and mesophiles, however certain composting detritivores such as black soldier fly larvae and redworms, also reduce many pathogens. The first stage of bokashi preserves the ingredients in a lactic acid fermentation. The acid is a natural disinfectant, used as such in household cleaning products, so that what enters the second (digestion) stage is essentially free of microbial pathogens. Thermophilic (high-temperature) composting is well known to destroy many seeds and nearly all types of pathogens (exceptions may include prions). The sanitizing qualities of (thermophilic) composting are desirable where there is a high likelihood of pathogens, such as with manure.

8.3 Materials that can be composted

8.3.1 Organic solid waste (Green waste)

As concern about landfill space increases, worldwide interest in recycling by means of composting is growing, since composting is a process for converting decomposable organic materials into useful stable products. Composting is one of the only ways to revitalize soil vitality due to phosphorus depletion in soil.

Co-composting is a technique that combines solid waste with dewatered biosolids, although difficulties in controlling inert and plastics contamination from municipal solid waste makes this approach less attractive.

Industrial composting systems are increasingly being installed as a waste management alternative to landfills, along with other advanced waste processing systems. Mechanical sorting of mixed waste streams combined with anaerobic digestion or in-vessel composting is called mechanical biological treatment, and is increasingly being used in developed countries due to regulations controlling the amount of organic matter allowed in landfills. Treatment of biodegradable waste before it enters a landfill reduces global warming from fugitive methane; untreated waste breaks down anaerobically in a landfill, producing landfill gas that contains methane, a potent greenhouse gas.

8.3.2 Animal manure and bedding

On many farms, the basic composting ingredients are animal manure generated on the farm and bedding. Straw and sawdust are common bedding materials. Non-traditional bedding materials are also used, including newspaper and chopped cardboard. The amount of manure composted on a livestock farm is often determined by cleaning schedules, land availability, and weather conditions. Each type of manure has its own physical, chemical, and biological characteristics. Cattle and horse manures, when mixed with bedding, possess good qualities for composting. Swine manure, which is very wet and usually not mixed with bedding material, must be mixed with straw or similar raw materials. Poultry manure also must be blended with carbonaceous materials—those low in nitrogen are preferred, such as sawdust or straw.

8.3.3 Human waste and sewage sludge

Human waste (excreta) can also be added as an input to the composting process since human waste is a nitrogen-rich organic material. It can be either composted directly, in composting toilets, or after mixing with water and treatment in a sewage treatment plant, in the form of sewage sludge treatment.

(1) Urine

People excrete far more water-soluble plant nutrients (nitrogen, phosphorus, potassium) in urine than in feces. Human urine can be used directly as fertilizer or put onto compost. Adding a healthy person's urine to compost usually will increase temperatures and therefore increase its ability to destroy pathogens and unwanted seeds. Unlike feces, urine does not attract disease-spreading flies (such as house flies or blow flies), and it does not contain the most hardy pathogens, such as parasite eggs. Urine usually does not smell for long, particularly when it is fresh, diluted, or put on sorbents.

(2) Humanure

"Humanure" is a portmanteau of human and manure, designating human excrement (faeces and urine) that is recycled via composting for agricultural or other purposes. The term was first used in a book by Joseph Jenkins in 1994 that advocates the use of this organic soil amendment. The term humanure is used by compost enthusiasts in the US but not generally elsewhere. Since the term "humanure" has no authoritative definition it is subject to various uses; news reporters occasionally fail to correctly distinguish between humanure and sewage sludge or "biosolids".

8.4 Uses

Compost can be used as an additive to soil, or other substrates such as coir and peat, as a soil improver, supplying humus and nutrients. It provides a rich growing medium, or a porous, absorbent material that holds moisture and soluble minerals, providing the support and nutrients in which plants can flourish, although it is rarely used alone, being primarily mixed with soil, sand, grit, bark chips, vermiculite, perlite, or clay granules to produce loam. Compost can be tilled directly into the soil or growing medium to increase the level of organic matter and the overall fertility of the soil. Compost that is ready to be used as an additive is dark brown or even black with an earthy smell.

Generally, direct seeding into a compost is not recommended due to the speed with which it may dry and the possible presence of phytotoxins that may inhibit germination, and the possible tie up of nitrogen by incompletely decomposed lignin. It is very common to see blends of $20\% \sim 30\%$ compost used for transplanting seedlings at cotyledon stage or later.

8.5 Composting technologies

Various approaches have been developed to handle different ingredients, locations, throughputs and applications for the composted product.

8.5.1 Industrial scale composting processes

Industrial scale composting can be carried out in the form of in-vessel composting, aerated static pile composting, vermicomposting, windrow composting and takes place in most Western countries now.

8.5.2 Vermicomposting

Vermicompost is the product or process of composting using various species of worms, usually red wigglers, white worms, and earthworms, to create a heterogeneous mixture of decomposed vegetable or food waste (excluding meat, dairy, fats, or oils), bedding materials, and vermicast. Vermicast, also known as worm castings, worm humus or worm manure, is the end product of the breakdown of organic matter by earthworms.

Vermicomposting is widely used in North America for on-site institutional processing of food waste, such as in hospitals, universities, shopping malls, and correctional facilities. Vermicomposting, also known as vermiculture, is used for medium-scale on-site institutional composting, such as for food waste from universities and shopping malls. It is selected either as a more environmentally friendly choice than conventional methods of disposal, or to reduce the cost of commercial waste removal.

Vermicomposting has gained popularity in both industrial and domestic settings, because as compared with conventional composting, it provides a way to compost organic materials more quickly (as defined by a higher rate of carbon-to-nitrogen ratio growth). It also generates products with lower salinity levels that are therefore more beneficial to plant mediums.

The earthworm species (or composting worms) most often used are red wigglers (*Eisenia fetida* or *Eisenia andrei*), though European nightcrawlers (*Eisenia hortensis* or *Dendrobaena veneta*) could also be used. Red wigglers are recommended by most vermiculture experts, as they have some of the best appetites and breed very quickly. Users refer to European nightcrawlers by a variety of other names, including dendrobaenas, dendras, Dutch nightcrawlers, and Belgian nightcrawlers.

Containing water-soluble nutrients, vermicompost is a nutrient-rich organic fertilizer and soil conditioner in a form that is relatively easy for plants to absorb. Worm castings are sometimes used as an organic fertilizer. As the earthworms grind and mix minerals uniformly in simple forms, plants can obtain them with minimal effort. The worms' digestive systems create environments that allow certain species of microbes to thrive to help create a "living" soil environment for plants. The fraction of soil which has gone through the digestive tract of earthworms is called the Drilosphere.

Researchers from the Pondicherry University discovered that worm composts can also be used to clean up heavy metals. The researchers found substantial reductions in heavy metals when the worms were released into the garbage and they were effective at removing lead, zinc, cadmium, copper and manganese.

8.5.3 Composting toilets

A composting toilet does not require water or electricity, and does not smell when properly managed. A composting toilet collects human excreta which is then added to a compost heap together with sawdust and straw or other carbon-rich materials, where pathogens are destroyed to some extent. The extent of pathogen destruction depends on the temperature (mesophilic or thermophilic conditions) and composting time. A composting toilet tries to process the excreta in situ although this is often coupled with a secondary external composting step. The resulting compost product has been given various names, such as humanure and Eco-

Humus.

A composting toilet can aid in the conservation of fresh water by avoiding the use of potable water required by the typical flush toilet. It further prevents the pollution of ground water by controlling the excreta decomposition before entering the system. When properly managed, there should be no ground water contamination from leachate.

8.5.4 Black soldier fly larvae composting

Black soldier fly (Hermetia illucens) larvae are able to rapidly consume large amounts of organic waste when kept at around 30℃. Black soldier fly larvae can reduce the dry matter of the organic waste by 73% and convert 16%~22% of the dry matter in the waste to biomass. The resulting compost still contains nutrients and can be used for biogas production, or further traditional composting or vermicomposting. The larvae are rich in fat and protein, and can be used as for example animal feed or biodiesel production. Enthusiasts have experimented with a large number of different waste products. Some even sell starter kits to the public. There are also larger-scale facilities.

8.5.5 Other systems at household level

(1) Raised garden beds or mounds

The practice of making raised garden beds or mounds filled with rotting wood is also called "Hügelkultur" in German. It is in effect creating a Nurse log that is covered with soil.

Benefits of raised garden beds include water retention and warming of soil. Buried wood becomes like a sponge as it decomposes, able to capture water and store it for later use by crops planted on top of the raised garden bed.

The buried decomposing wood will also give off heat, as all compost does, for several years. These effects have been used by Sepp Holzer to enable fruit trees to survive at otherwise inhospitable temperatures and altitudes.

(2) Bokashi

Bokashi composting is a method of covering food waste or wilted plants with a mixture of microorganisms to decrease smell, reduce the risk of attracting pests and increase the rate of decomposition. Bokashi is Japanese for "shading off" or "gradation". It derives from the practice of Japanese farmers centuries ago of covering food waste with rich, local soil that contained the microorganisms that would ferment the waste.

The technique relies on effective microorganisms. These essential microbes are typically added to the food waste using an inoculated bokashi bran.

Newspaper fermented in a lactobacillus culture can substitute for bokashi bran for a successful bokashi bucket.

(3) Compost tea

Compost teas are defined as water extracts brewed from composted materials and can be derived from aerobic or anaerobic processes. Compost teas are generally produced from adding one volume of compost to $4 \sim 10$ volumes of water, but there has also been debate about the benefits of aerating the mixture. Field studies have shown the benefits of addition compost teas to crops due to the addition of organic matter, increased nutrient availability and increased microbial activity. They have also been shown to have an effect on plant pathogens.

8.5.6 Related technologies

Anaerobic digestion is process for converting organic waste into biogas. The residual material, sometimes in combination with sewage sludge, can be treated with an aerobic composting process before the compost is sold or given away.

8.6 Regulations

There have been process and product guidelines in Europe since the early 1980s (Germany, the Netherlands, Switzerland) while only more recently in the UK and the US. In both these countries, private trade associations within the industry have established loose standards, some say as a stop-gap measure to discourage independent government agencies from establishing tougher consumer-friendly standards.

The USA is the only Western country that does not distinguish sludge-source compost from green compost, and by default 50% of states in the USA expect composts to comply in some manner with the Federal EPA 503 Rule promulgated in 1984 for sludge products.

Compost is regulated in Canada and Australia as well.

Many countries and some individual cities such as Seattle and San Francisco require food and yard waste to be sorted for composting (Mandatory Recycling and Composting Ordinance in San Francisco).

8.7 Examples

Large-scale composting systems are used by many urban areas around the world.

The world's largest municipal solid waste co-composter is the Edmonton Composting Facility in Edmonton, Alberta, Canada, which turns 220,000 of residential solid waste and 22,500 of dry biosolids per year into 80,000 t of compost. The facility is 38,690 m^2 (416,500 ft^2) in area, equivalent to $4\frac{1}{2}$ Canadian football fields, and the operating structure is the largest stainless steel building in North America, with the size of 14 NHL rinks.

In 2006, Qatar awarded Keppel Seghers Singapore, a subsidiary of Keppel Corporation, a contract to begin construction on a 275,000 t/a anaerobic digestion and composting plant licensed by Kompogas Switzerland. This plant, with 15 independent anaerobic digesters, would be the world's largest composting facility once fully operational in early 2011 and formed part of Qatar's Domestic Solid Waste Management Centre, the largest integrated waste management complex in the Middle East.

Another large MSW composter is the Lahore Composting Facility in

Lahore, Pakistan, which has a capacity to convert 1,000 t of MSW per day into compost. It also has a capacity to convert substantial portion of the intake into refuse-derived fuel (RDF) materials for further combustion use in several energy-consuming industries across Pakistan, for example in cement manufacturing companies where it is used to heat cement kilns. This project has also been approved by the Executive Board of the United Nations Framework Convention on Climate Change for reducing methane emissions, and has been registered with a capacity of reducing 108,686 t of carbon dioxide equivalent per year.

Kew Gardens in London has one of the biggest non-commercial compost heaps in Europe.

8.8 History

Composting as a recognized practice dates to at least the early Roman Empire, and was mentioned as early as Cato the Elder's 160 BCE piece De Agri Cultura. Traditionally, composting involved piling organic materials until the next planting season, when the materials would have decayed enough to be ready for use in the soil. The advantage of this method is that little working time or effort is required from the composter and it fits in naturally with agricultural practices in temperate climates. Disadvantages (from the modern perspective) are that space is occupied for a whole year, and some nutrients might be leached due to the exposure to rainfall, and disease-producing organisms and insects may not be adequately controlled.

Composting was somewhat modernized at the beginning of the 1920s in Europe as a tool for organic farming. The first industrial station for the transformation of urban organic materials into compost was set up in Wels, Austria in 1921. Early frequent citations for propounding composting within farming are Rudolf Steiner for the German-speaking world, the founder of a farming method called biodynamics, and Annie Francé-Harrar, who was appointed on behalf of the government in Mexico and supported the country to set up a large humus organization in the fight against erosion and soil degradation from 1950 to 1958.

In the English-speaking world, it was Sir Albert Howard who worked extensively in India on sustainable practices and Lady Eve Balfour who was a big proponent of composting. Composting was imported to America by various followers of these early European movements, such as J. I. Rodale (founder of Rodale Organic Gardening), E. E. Pfeiffer (who developed scientific practices in biodynamic farming), Paul Keene (founder of Walnut Acres in Pennsylvania), and Scott and Helen Nearing (who inspired the Back-to-the-Land Movement of the 1960s). Coincidentally, some of the above met briefly in India-all were quite influential in the U. S. from the 1960s into the 1980s.

Chapter 9 Life cycle assessment (LCA) for municipal solid waste management

Over their life-time, products (goods and services) can contribute to various environmental impacts. Life cycle thinking considers the range of impacts throughout the life of a product, as shown in Fig. 32. Life cycle assessment quantifies this by assessing the emissions, resources consumed and pressures on health and the environment that can be attributed to a product. It takes the entire life cycle into account—from the extraction of natural resources to material processing, manufacturing, distribution and use; and finally the re-use, recycling, energy recovery and the disposal of remaining waste.

The fundamental aim of Life cycle thinking is to reduce overall environmental impacts. This may involve trade-offs between impacts at different stages of the life cycle. However, care needs to be taken to avoid shifting problems from one stage to another. Reducing the environmental impact of a product at the production stage may lead to a greater environmental impact further down the line. An apparent benefit of a waste management option can therefore be cancelled out if not thoroughly evaluated. The European Commission has developed guidelines for Life Cycle Assessment which are fully compatible with international standards. These aim to ensure quality and consistency based on scientific evidence when carrying out assessments.

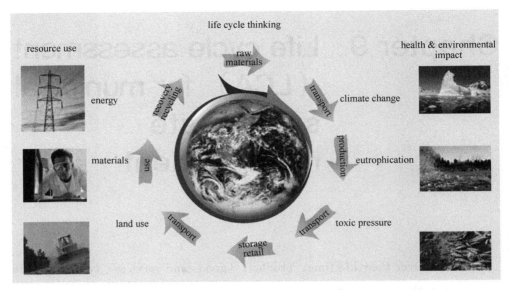

Fig. 32 Schematic of life cycle thinking

9.1 Introduction—What is LCA and how is it useful?

Life cycle assessment (LCA, also known as life cycle analysis, ecobalance, and cradle-to-grave analysis) is a technique to assess environmental impacts associated with all the stages of a product's life from raw material extraction through materials processing, manufacture, distribution, use, repair and maintenance, and disposal or recycling. Designers use this process to help critique their products. LCAs can help avoid a narrow outlook on environmental concerns by: (1) Compiling an inventory of relevant energy and material inputs and environmental releases; (2) Evaluating the potential impacts associated with identified inputs and releases; (3) Interpreting the results to help make a more informed decision.

When materials are recycled they are made available for use for several future life cycles and can therefore replace virgin material more than just once. It is important to distinguish this approach for material flow analysis for a given material from life cycle analysis of products. A product life cycle analysis analyses the product system from cradle to grave, but uses some form of allocation in order to separate the life cycle of one product from another in the case where component materials are recycled. The focus here is the flow of the given material from cradle (raw material extraction) to grave (the material, or its inherent energy, is no longer available for use). The limitation on the number of times materials can be recycled is set by either the recycling rate, or the technical properties of the recycled material.

9.2 Goals and purpose of LCA

The goal of LCA is to compare the full range of environmental effects assignable to products and services by quantifying all inputs and outputs of material flows and assessing how these material flows affect the environment. This information is used to improve processes, support policy and provide a sound basis for informed decisions.

The term life cycle refers to the notion that a fair, holistic assessment requires the assessment of raw-material production, manufacture, distribution, use and disposal, including all intervening transportation steps necessary or caused by the product's existence.

There are two main types of LCA. Attributional LCAs seek to establish (or attribute) the burdens associated with the production and use of a product, or with a specific service or process, at a point in time (typically the recent past). Consequential LCAs seek to identify the environmental consequences of a decision or a proposed change in a system under study (oriented to the future), which means that market and economic implications of a decision may have to be taken into account. Social LCA is under development as a different approach to life cycle thinking intended to assess social implications or potential impacts. Social LCA should be considered as an approach that is complementary to environmental LCA.

The procedures of life cycle assessment are part of the ISO 14000 environmental management standards: in ISO 14040: 2006 and 14044:

2006. (ISO 14044 replaced earlier versions of ISO 14041 to ISO 14043.) GHG product life cycle assessments can also comply with specifications such as PAS 2050 and the GHG Protocol Life Cycle Accounting and Reporting Standard.

9.3 Four main phases

According to the ISO 14040 and 14044 standards, a life cycle assessment is carried out in four distinct phases as illustrated in Fig. 33. The phases are often interdependent in that the results of one phase will inform how other phases are completed.

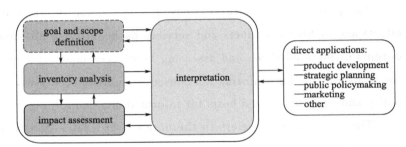

Fig. 33　Elements of life cycle assessment

9.3.1 Goal and scope

An LCA starts with an explicit statement of the goal and scope of the study, which sets out the context of the study and explains how and to whom the results are to be communicated. This is a key step and the ISO standards require that the goal and scope of an LCA be clearly defined and consistent with the intended application. The goal and scope document therefore includes technical details that guide subsequent work: (1) The functional unit, which defines what precisely is being studied and quantifies the service delivered by the product system, providing a reference to which the inputs and outputs can be related. Further, the functional unit is an important basis that enables alternative goods, or services, to be compared and analyzed. So to explain this, a functional system that could be inputs, processes and outputs contains a functional unit, which fulfills a function, for example paint is covering a wall, and making a functional unit of 1m^2 covered for 10 years. The functional flow would be the items necessary for that function, so it would be a brush, tin of paint and the paint itself. (2) The system boundaries, which are definitions of which processes should be included in the analysis of a product system. (3) Any assumptions and limitations. (4) The allocation methods used to partition the environmental load of a process when several products or functions share the same process; allocation is commonly dealt with in one of three ways: system expansion, substitution and partition. Doing this is not easy and different methods may give different results. (5) The impact categories chosen, for example human toxicity, smog, global warming, eutrophication.

9.3.2 Life cycle inventory

Life cycle inventory (LCI) analysis involves creating an inventory of flows from and to nature for a product system. Inventory flows include inputs of water, energy, and raw materials, and releases to air, land and water. To develop the inventory, a flow model of the technical system is constructed using data on inputs and outputs. The flow model is typically illustrated with a flow chart that includes the activities that are going to be assessed in the relevant supply chain and gives a clear picture of the technical system boundaries. The input and output data needed for the construction of the model are collected for all activities within the system boundary, including from the supply chain (referred to as inputs from the technosphere).

The data must be related to the functional unit defined in the goal and scope definition. Data can be presented in tables and some interpretations can be made already at this stage. The results of the inventory is an LCI which provides information about all inputs and outputs in the form of elementary flow to and from the environment from all the unit processes involved in the study.

Inventory flows can number in the hundreds depending on the system boundary. For product LCAs at either the generic (i. e., representative industry averages) or brand-specific level, that data is typically collected through survey questionnaires. At an industry level, care has to be taken to ensure that questionnaires are completed by a representative sample of producers, leaning toward neither the best nor the worst, and fully representing any regional differences due to energy use, material sourcing or other factors. The questionnaires cover the full range of inputs and outputs, typically aiming to account for 99% of the mass of a product, 99% of the energy used in its production and any environmentally sensitive flows, even if they fall within the 1% level of inputs.

One area where data access is likely to be difficult is flows from the technosphere. The technosphere is more simply defined as the man-made world. Considered by geologists as secondary resources, these resources are in theory 100% recyclable; however, in a practical sense, the primary goal is salvage. For an LCI, these technosphere products (supply chain products) are those that have been produced by man and unfortunately those completing a questionnaire about a process which uses a man-made product as a means to an end will be unable to specify how much of a given input they use. Typically, they will not have access to data concerning inputs and outputs for previous production processes of the product. The entity undertaking the LCA must then turn to secondary sources if it does not already have that data from its own previous studies. National databases or data sets that come with LCA-practitioner tools, or that can be readily accessed, are the usual sources for that information. Care must then be taken to ensure that the secondary data source properly reflects regional or national conditions.

9.3.3 Life cycle impact assessment

Inventory analysis is followed by impact assessment. This phase of LCA is aimed at evaluating the significance of potential environmental impacts based on the LCI flow results. Classical life cycle impact assessment (LCIA) consists of the following mandatory elements: (1) selection of impact categories, category indicators, and characterization models; (2) the classification stage, where the inventory parameters are sorted and assigned to specific impact categories; (3) impact measurement, where the categorized LCI flows are characterized, using one of many possible LCIA methodologies, into common equivalence units that are then summed to provide an overall impact category total.

In many LCAs, characterization concludes the LCIA analysis; this is also the last compulsory stage according to ISO 14044: 2006. However, in addition to the above mandatory LCIA steps, other optional LCIA elements (normalization, grouping, and weighting) may be conducted depending on the goal and scope of the LCA study. In normalization, the results of the impact categories from the study are usually compared with the total impacts in the region of interest, the U. S. for example. Grouping consists of sorting and possibly ranking the impact categories. During weighting, the different environmental impacts are weighted relative to each other so that they can then be summed to get a single number for the total environmental impact. ISO 14044: 2006 generally advises against weighting, stating that "weighting, shall not be used in LCA studies intended to be used in comparative assertions intended to be disclosed to the public". This advice is often ignored, resulting in comparisons that can reflect a high degree of subjectivity as a result of weighting.

Life cycle impacts can also be categorized under the several phases of the development, production, use, and disposal of a product. Broadly speaking, these impacts can be divided into "First Impacts", use impacts, and end of life impacts. "First Impacts" include extraction of raw materials, manufacturing (conversion of raw materials into a product), transportation of the product to a market or site, construction/installation, and the beginning of the use or occupancy. Use impacts include physical impacts of operating the product or facility (such as energy, water, etc.), maintenance, renovation and repairs (required to continue to use the product or facility). End of life impacts include demolition and processing of waste or recyclable materials.

9.3.4 Interpretation

Life cycle interpretation is a systematic technique to identify, quantify, check, and evaluate information from the results of the life cycle inventory and/or the life cycle impact assessment. The results from the inventory analysis and impact assessment are summarized during the interpretation phase. The outcome of the interpretation phase is a set of conclusions and recommendations for the study. According to ISO 14040: 2006, the interpretation should include: (1) identification of significant issues based on the results of the LCI and LCIA phases of an LCA; (2) evaluation of the study considering completeness, sensitivity and consistency checks; (3) conclusions, limitations and recommendations.

A key purpose of performing life cycle interpretation is to determine the level of confidence in the final results and communicate them in a fair, complete, and accurate manner. Interpreting the results of an LCA is not as simple as "3 is better than 2, therefore Alternative A is the best choice"! Interpreting the results of an LCA starts with understanding the accuracy of the results, and ensuring they meet the goal of the study. This is accomplished by identifying the data elements that contribute significantly to each impact category, evaluating the sensitivity of these significant data elements, assessing the completeness and consistency of the study, and drawing conclusions and recommendations based on a clear understanding of how the LCA was conducted and the results were developed.

9.4 LCA uses

Based on a survey of LCA practitioners carried out in 2006, LCA is mostly used to support business strategy (18%) and R&D (18%), as input to product or process design (15%), in education (13%) and for labeling or product declarations (11%). LCA will be continuously integrated into the built environment as tools such as the European ENSLIC Building project guidelines for buildings or developed and implemented, which provide practitioners guidance on methods to implement LCI data into the planning and design process.

Major corporations all over the world are either undertaking LCA in house or commissioning studies, while governments support the development of national databases to support LCA. Of particular note is the growing use of LCA for ISO Type III labels called Environmental Product Declarations, defined as "quantified environmental data for a product with pre-set categories of parameters based on the ISO 14040 series of standards, but not excluding additional environmental information". These third-party certified LCA-based labels provide an increasingly important basis for assessing the relative environmental merits of competing products. Third-party certification plays a major role in today's industry. Independent certification can show a company's dedication to safer and environmental friendlier products to customers and NGOs.

LCA also has major roles in environmental impact assessment, integrated waste management and pollution studies. A recent study evaluated the LCA of a laboratory-scale plant for oxygen-enriched air production coupled with its economic evaluation in a holistic eco-design standpoint.

9.5 Data analysis

A life cycle analysis is only as valid as its data; therefore, it is crucial that data used for the completion of a life cycle analysis are accurate and current. When comparing different life cycle analyses with one another, it is crucial that equivalent data are available for both products or processes in question. If one product has a much higher availability of data, it cannot be justly compared to another product which has less detailed data.

There are two basic types of LCA data-unit process data and environmental input-output data (EIO), where the latter is based on national economic input-output data. Unit process data are derived from direct surveys of companies or plants producing the product of interest, carried out at a unit process level defined by the system boundaries for the study.

Data validity is an ongoing concern for life cycle analyses. Due to globalization and the rapid pace of research and development, new materials and manufacturing methods are continually being introduced to the market. This makes it both very important and very difficult to use up-to-date information when performing an LCA. If an LCA's conclusions are to be valid, the data must be recent; however, the data-gathering process takes time. If a product and its related processes have not undergone significant revisions since the last LCA data was collected, data validity is not a problem. However, consumer electronics such as cell phones can be redesigned as often as every 9 to 12 months, creating a need for ongoing data collection.

The life cycle considered usually consists of a number of stages including: materials extraction, processing and manufacturing, product use, and product disposal. If the most environmentally harmful of these stages can be determined, then impact on the environment can be efficiently reduced by focusing on making changes for that particular phase. For example, the most energy-intensive life phase of an airplane or car is during use due to fuel consumption. One of the most effective ways to increase fuel efficiency is to decrease vehicle weight, and thus, car and airplane manufacturers can decrease environmental impact in a significant way by replacing heavier materials with lighter ones such as aluminium or carbon fiber-reinforced elements. The reduction during the use phase should be more than enough to balance additional raw material or manufacturing cost.

9.6 Variants

9.6.1 Cradle-to-grave

Cradle-to-grave is the full Life cycle assessment from resource extraction ("cradle") to use phase and disposal phase ("grave"). For example, trees produce paper, which can be recycled into low-energy production cellulose (fibered paper) insulation material, then used as an energy-saving device in the ceiling of a home for 40 years, saving 2,000 times the fossil-fuel energy used in its production. After 40 years the cellulose fibers are replaced and the old fibers are disposed of, possibly incinerated. All inputs and outputs are considered for all the phases of the life cycle.

9.6.2 Cradle-to-gate

Cradle-to-gate is an assessment of a partial product life cycle from resource extraction (cradle) to the factory gate (i.e., before it is transported to the consumer). The use phase and disposal phase of the product are omitted in this case. Cradle-to-gate assessments are sometimes the basis for environmental product declarations (EPD) termed business-to-business EDPs. One of the significant uses of the cradle-to-gate approach is compiling the life cycle inventory (LCI). This allows the LCA to collect all of the impacts leading up to resources being purchased by the facility. They can then add the steps involved in their transport to plant and manufacture process to produce their own cradle-to-gate values more easily for their products.

9.6.3 Cradle-to-cradle or closed loop production

Cradle-to-cradle is a specific kind of cradle-to-grave assessment, where the end-of-life disposal step for the product is a recycling process. It is a method used to minimize the environmental impact of products by employing sustainable production, operation, and disposal practices and aims to incorporate social responsibility into product development. From the recycling process originate new, identical products (e.g., asphalt pavement from discarded asphalt pavement, glass bottles from collected glass bottles), or different products (e.g., glass wool insulation from collected glass bottles).

Allocation of burden for products in open loop production systems presents considerable challenges for LCA. Various methods, such as the avoided burden approach have been proposed to deal with the issues involved.

9.6.4 Gate-to-gate

Gate-to-gate is a partial LCA looking at only one value-added process in the entire production chain. Gate-to-gate modules may also later be linked in their appropriate production chain to form a complete cradle-to-gate evaluation.

9.6.5 Well-to-wheel

Well-to-wheel is the specific LCA used for the transport of fuels and vehicles. The analysis is often broken down into stages entitled "well-to-station" or "well-to-tank", and "station-to-wheel" or "tank-to-wheel", or "plug-to-wheel". The first stage, which incorporates the feedstock or fuel production and processing and fuel delivery or energy transmission, is called the "upstream" stage, while the stage that deals with vehicle operation itself is sometimes called the "downstream" stage. The well-to-wheel analysis is commonly used to assess total energy consumption, or the energy conversion efficiency and emissions impact of marine vessels, aircrafts and motor vehicles, including their carbon footprint, and the fuels used in each of these transport modes. Well-to-wheel (WTW) analysis is useful for reflecting the different efficiencies and emissions of energy technologies and fuels at both the upstream and downstream stages, giving a more complete picture of real emissions.

The well-to-wheel variant has a significant input on a model developed by Argonne National Laboratory. The Greenhouse Gases, Regulated Emissions, and Energy Use in Transportation Model was developed to evaluate the impacts of new fuels and vehicle technologies. The model evaluates the impacts of fuel use using a well-to-wheel evaluation while a traditional cradle-to-grave approach is used to determine the impacts from the vehicle itself. The model reports energy use, greenhouse gas emissions, and six additional pollutants: volatile organic compounds (VOCs), carbon monoxide (CO), nitrogen oxide (NO_x), particulate matter with size smaller than 10 μm (PM_{10}), particulate matter with size smaller than 2.5 μm ($PM_{2.5}$), and sulfur oxides (SO_x).

Quantitative values of greenhouse gas emissions calculated with the WTW or with the LCA method can differ, since the LCA is considering more emission sources. For example, while assessing the GHG emissions of a battery electric vehicle in comparison with a conventional internal combustion engine vehicle, the WTW (accounting only the GHG for manufacturing the fuels) finds out that an electric vehicle can save 50%~60% of the GHG, while an hybrid LCA-WTW method, considering the GHG due to both the manufacturing and the end of life of the battery, gives GHG emission savings 10%~13% lower, compared to the WTW.

9.6.6 Economic input-output life cycle assessment

Economic input-output LCA (EIOLCA) involves use of aggregate sector-level data on how much environmental impact can be attributed to each sector of the economy and how much each sector purchases from other sectors. Such analysis can account for long chains (for example, manufacturing an automobile requires energy, but producing energy requires equipments, and manufacturing those equipments requires energy, etc.), which somewhat alleviates the scoping problem of conventional LCA; however, EIOLCA relies on sector-level averages that may or may not be representative of the specific subset of the sector relevant to a particular product and therefore is not suitable for evaluating the environmental impacts of products. Additionally the conversion of economic volume into environmental impacts has not been validated.

9.6.7 Ecologically based LCA

While a conventional LCA uses many of the same approaches and strategies as an Eco-LCA, the latter considers a much broader range of ecological impacts. It was designed to provide a guide to wise management of human activities by understanding the direct and indirect impacts on ecological resources and surrounding ecosystems. Developed by the Ohio State University Center for Resilience, Eco-LCA is a methodology that quantitatively takes into account regulating and supporting services during the life cycle of economic goods and products. In this approach services are categorized in four main groups: supporting, regulating, provisioning and cultural services.

9.7 Exergy based LCA

Exergy of a system is the maximum useful work possible during a process that brings the system into equilibrium with a heat reservoir. Wall clearly states the relation between exergy analysis and resource accounting. This intuition confirmed by DeWulf and Sciubba lead to exergo-economic accounting and to methods specifically dedicated to LCA such as exergetic material input per unit of service (EMIPS). The concept of material input per unit of service (MIPS) is quantified in terms of the second law of thermodynamics, allowing the calculation of both resource input and service output in exergy terms. This exergetic material input per unit of service (EMIPS) has been elaborated for transport technology. The service not only takes into account the total mass to be transported and the total distance, but also the mass per single transport and the delivery time.

9.8 Life cycle energy analysis

Life cycle energy analysis (LCEA) is an approach in which all energy inputs to a product are accounted for, not only direct energy inputs during manufacture, but also all energy inputs needed to produce components, materials and services needed for the manufacturing process. An earlier term for the approach was energy analysis. With LCEA, the total life cycle energy input is established.

9.8.1 Energy production

It is recognized that much energy is lost in the production of energy commodities themselves, such as nuclear energy, photovoltaic electricity or high-quality petroleum products. Net energy content is the energy content of the product minus energy input used during extraction and conversion, directly or indirectly. A controversial early result of LCEA claimed that manufacturing solar cells requires more energy than can be recovered in using the solar cell. The result was refuted. Another new concept from life cycle assessments is Energy Cannibalism. Energy Cannibalism refers to an effect where rapid growth of an entire energy-intensive industry creates a need for energy that uses (or cannibalizes) the energy of existing power plants. Thus during rapid growth the industry as a whole produces no energy because new energy is used to fuel the embodied energy of future power plants. Work has been undertaken in the UK to determine the life cycle energy (alongside full LCA) impacts of a number of renewable technologies.

9.8.2 Energy recovery

If materials are incinerated during the disposal process, the energy released during burning can be harnessed and used for electricity production. This provides a low-impact energy source, especially when compared with coal and natural gas. While incineration produces more greenhouse gas emissions than landfilling, the waste plants are well-fitted with filters to minimize this negative impact. A recent study comparing energy consumption and greenhouse gas emissions from landfilling (without energy recovery) against incineration (with energy recovery) found incineration to be superior in all cases except for when landfill gas is recovered for electricity production.

9.8.3 Criticism

A criticism of LCEA is that it attempts to eliminate monetary cost analysis; that is replacing the currency by which economic decisions are made with an energy currency. It has also been argued that energy efficiency is only one consideration in deciding which alternative process to employ, and that it should not be elevated to the only criterion for determining environmental acceptability; for example, simple energy analysis does not take into account the renewability of energy flows or the toxicity of waste products; however the life cycle assessment does help companies become more familiar with environmental properties and improve their environmental system. Incorporating Dynamic LCAs of renewable energy technologies (using sensitivity analyses to project future improvements in renewable systems and their share of the power grid) may help mitigate this criticism.

In recent years, the literatures on life cycle assessment of energy technologies have begun to reflect the interactions between the current power grid and future energy technology. Some papers have focused on energy life cycle, while others have focused on carbon dioxide (CO_2) and other greenhouse gases. The essential critique given by these sources is that when considering energy technology, the growing nature of the power grid must be taken into consideration. If this is not done, a given class of energy technology may emit more CO_2 over its lifetime than it mitigates.

A problem the energy analysis method cannot resolve is that different energy forms (heat, electricity, chemical energy, etc.) have different quality and value even in natural sciences, as a consequence of the two main laws of thermodynamics. A thermodynamic measure of the quality of energy is exergy. According to the first law of thermodynamics, all energy inputs should be accounted with equal weight, whereas by the second law diverse energy forms should be accounted with different values.

The conflict is resolved in one of these ways: (1) value difference between energy inputs is ignored; (2) a value ratio is arbitrarily assigned (e. g. , a Joule of electricity is 2.6 times more valuable than a Joule of heat or fuel input); (3) the analysis is supplemented by economic (monetary) cost analysis; (4) exergy instead of energy can be the metric used for the life cycle analysis.

9.9 Critiques

Life cycle assessment is a powerful tool for analyzing commensurable aspects of quantifiable systems. Not every factor, however, can be reduced to a number and inserted into a model. Rigid system boundaries make it difficult to account for changes in the system. This is sometimes referred to as the boundary critique to systems thinking. The accuracy and availability of data can also contribute to inaccuracy. For instance, data from generic processes may be based on averages, unrepresentative sampling, or outdated results. Additionally, social implications of products are generally absent in LCAs. Comparative life cycle analysis is often used to determine a better process or product to use. However, because of aspects like different system boundaries, different statistical information, different product uses, etc. , these studies can easily be swayed in favor of one product or process over another in one study and the opposite in another study based on varying parameters and different available data. There are guidelines to help reduce such conflicts in results but the method still provides a lot of room for the researchers to decide what is important, how the product is typically manufactured, and how it is typically used.

An in-depth review of 13 LCA studies of wood and paper products found a lack of consistency in the methods and assumptions used to track carbon during the product lifecycle. A wide variety of methods and assumptions were used, leading to different and potentially contrary conclusions-particularly with regard to carbon sequestration and methane generation in landfills and carbon accounting during forest growth and product use.

Streamline LCA

This process includes three steps. First, a proper method should be selected to combine adequate accuracy with acceptable cost burden in order to guide decision making. Actually, in LCA process, besides streamline LCA, Eco-screening and complete LCA are usually considered as well. However, the former one could only provide limited details and the latter one with more detailed information is more expensive. Second, single measure of stress should be selected. Typical LCA output includes resource consumption, energy consumption, water consumption, emission of CO_2, toxic residues and so on. One of these outputs is used as the main factor to measure in streamline LCA. Energy consumption and CO_2 emission are often regarded as "practical indicators". Last, stress selected in step 2 is used as standard to assess phase of life separately and identify the most damaging phase. For instance, for a family car, energy consumption could be used as the single stress factor to assess each phase of life. The result shows that the most energy intensive phase for a family car is the usage stage.

Life cycle assessment of engineered material in service plays a significant role in saving energy, conserving resources and saving cost by preventing premature failure of critical engineered component in a machine or equipment. LCA data of surface engineered materials are used to improve life cycle of the engineered component. Life cycle improvement of industrial machineries and equipments including manufacturing, power generation, transportations, etc. leads to improvement in energy efficiency, sustainability and negating global temperature rise. Estimated reduction in anthropogenic carbon emission is minimum 10% of the global emission.

9.10　Supporting waste management decisions-Examples

European, national and local public authorities and businesses are increasingly being encouraged to make use of life cycle assessment as supporting tools for decisionmaking.

Waste management is an area where local conditions often influence the choice of policy options. Life cycle thinking and life cycle assessment can be used to weigh up the possible environmental benefits and drawbacks linked to policy options in a specific situation.

Typical questions that can arise in local or regional settings include: (1) Is it better to recycle waste or to recover energy from it? What are the trade-offs for particular waste streams? (2) Is it better to replace appliances with new, more energy efficient models or keep using the old ones and avoid generating waste? (3) Are the greenhouse gas emissions created when collecting waste offset by the expected benefits?

Life cycle thinking and assessment can be used to support decision-making in the area of waste management and to identify the best environmental options. It can help policy makers understand the benefits and trade-offs they have to face when making decisions on waste management strategies. It gives quantitative information which puts potential environmental advantages and disadvantages into perspective. Life cycle assessment cannot replace a decision-making process but it can guide public authorities and businesses to make better environmental choices. It should be noted that the examples given here are valid for their specific situation and their conclusions cannot be generalised.

References

[1] Kumar S, Dhar H, Vijay V, et al. Characterization of municipal solid waste in high altitude sub-tropical regions [J]. Environmental Technology, 2016: 1-27.

[2] Herbert I. Centenary history of waste and waste managers in London and south east England [R]. London: Chartered Institution of Wastes Management (CIWM), 2008.

[3] U. S. Environmental Protection Agency. Non-hazardous waste-municipal solid waste [R]. Office of Brownfields, Tetra Tech, Em Inc., 2016.

[4] U. S. Energy Information Administration. Municipal Solid Waste [R]. Office of Coal, Nuclear, Electric and Alternate Fuels U. S. Department of Energy, 2007.

[5] Mechanical Biological Treatment Welsh Assembly (MBTWA). "Mechanical biological treatment", Environment Countryside and Planning Website, Welsh Assembly, 2005: 21-32.

[6] Christchurch City Council. Organics -Green Bin [R]. New Zealand Government, 2016.

[7] UNT Digital Library. Advancing sustainable materials management: Facts and figures [R]. University of North Texas, 2013.

[8] Heather R. Gone tomorrow: The hidden life of garbage [J]. Library Journal, 2005: 78.

[9] U. S. EPA Combustion emissions from hazardous waste incinerators, boilers and industrial furnaces, and municipal solid waste incinerators- Results from five STAR grants and research needs [R]. U. S. EPA, 2016.

[10] Zhang D Q, Tan S K, Gersberg R M, et al. Municipal solid waste management in China: status, problems and challenges [J]. Journal of environmental management, 2010: 1623-1633.

[11] Brunner P H, Ernst W R. Alternative methods for the analysis of municipal solid waste [J]. Waste Management & Research, 1986: 147-160.

[12] Postel S. Carrying capacity: Earth's bottom line [J]. Challenge, 1994, 37 (2): 4-12.

[13] Poulsen T G. Landfilling, past, present and future [J]. Waste Management Research, 2014: 177-178.

[14] Wong J W C, Selvam A. Waste management in Hong Kong: A lesson to learn [C]. Proc., 4th International Conference on Solid Waste Management and Exhibition on Municipal Services, Urban Development & Clean Technology, Acharya NGR Agriculture University, Hyderabad, India, 2014.

[15] Li Z, Fu H, Qu X. Estimating municipal solid waste generation by different activities and various resident groups: A case study of Beijing [J]. Science of Total Environment, 2011: 4406-4414.

[16] Gomez G, Meneses M, Ballinas L, et al. Seasonal characterization of municipal solid waste (MSW) in the city of Chihuahua, Mexico [J]. Waste Management, 2009: 2018-2024.

[17] Wakadikar K, Sil A, Kumar S, et al. Influence of sewage sludge and leachate on biochemical methane potential of waste biomass [J]. Journal of Bioremediation & Biodegra-

dation, 2012: S8.

[18] Hoornweg D, Bhada-Tata P. What a waste: A global review of solid waste management, urban development series [C]. World Bank, Washington DC, 2012.

[19] EB (Environment Bureau) of Hong Kong. Hong Kong blueprint for sustainable use of resources 2013-2022 [R]. Hong Kong Special Administrative Region, Hong Kong, 2013.

[20] Das A, Gupta A K, Mazumder T N. Vulnerability assessment using hazard potency for regions generating industrial hazardous waste [J]. Journal of Hazardous Materials, 2012: 308-317.

[21] U. S. Environmental Protection Agency. Solid waste and emergency response (5305W), EPA530-K-05-012 [C]. Washington, DC, 2005.

[22] Marinkovic N, Vitale L, Holcer N J, et al. Management of hazardous medical waste in Croatia [J]. Waste Management, 2008: 1049-1056.

[23] Tasaki T, Hashimoto S, Moriguchi Y. A quantitative method to evaluate the level of material use in lease/reuse systems of electricity and electronic equipment [J]. Journal of Cleaner Production, 2006: 1519-1528.

[24] Lim S R, Schoenung J M. Human health and ecological toxicity potentials due to heavy metal content in waste electronic devices with flat panel displays [J]. Journal of Hazardous Materials, 2010: 251-259.

[25] Bouvier R, Wagner T. The influence of collection facility attributes on household collection rates of electronic waste: The case of televisions and computer monitors [J]. Resources Conservation Recycling, 2011: 1051-1059.

[26] UNEP IETC (United Nations Environment Programme International Environmental Technology Center). International source book on environmentally sound technologies for municipal solid waste management [R]. Washington, DC, 1996.

[27] EC (European Commission). Preparing a waste prevention programme, guidance document [R]. European Commission DG Environment, Brussels, Belgium, 2012.

[28] OECD. Working group on waste prevention and recycling - Working group on environmental information and outlooks: OECD workshop on waste prevention - Toward Performance Indicators [R]. 2002.

[29] EEA (European Environment Agency). Case studies on waste minimization practices in Europe [R]. Topic Rep. , Copenhagen, 2002.

[30] DEFRA. Household waste prevention evidence review [R]. Final Rep. WR1204, Brook Lyndhurst, London, 2009.

[31] EC (European Commission). Preparatory study on food waste across EU-27 [R]. Technical Rep. 2010-054, Brussels, Belgium, 2011.